浓浓亚洲风
——传承与演绎东方古韵的

黄滢 主编

江苏科学技术出版社

超越豪宅

看得到的是细节，看不到的是一切

岩舍国际设计事务所　林济民

我是老子学说的信徒，毕生追求"无为而治"的境界，但有趣的是，我每天努力不懈地要求自己求新求变，也要求员工专注谨慎。为此，本人现在的从业心态倒比较像个"布道家"，与自己的"无为而治"渐行渐远。

设计师是近代新兴的行业，为消费者提供新的想法和新的便利，打造新的环境和新的生活，但是所有新想法却是基于最深层和最基本的个人特质、修养、生活经历和所处的社会环境。以前，全世界的设计师都无法超越他所处的环境做深度设计，但是这些年已不然，深入人心的"地球村"概念、交通的便利和旅行的普及渐渐地改变了这一切。这些改变使个人突破了因环境所限而带来的思维局限，而设计师的作品则更多地体现出他在文学、美术、音乐等人文艺术方面的综合素养和底蕴，体现出他个人的气度和品性。

豪宅定义

何谓"豪宅"？答案众说纷纭，这个词在房地产销售中已被滥用，但是总体而言应该是指面积超过300 m^2的单层建筑物。目前，新的豪宅代名词是"招待会馆"或"招待会所"，意思是家大得可以招待多位亲友入住，如同饭店般舒适、大气。每个房间都有2 m × 2.3 m的大床和进口卫浴设备，客厅可容纳10人以上，厨房是中央系统，餐厅也能供8人以上同时用餐，当然，最好要有起居室、书房、视听室、佣人房及次客厅和次餐厅等功能空间。拥有以上规模的建筑物才能称得上是豪宅，但是本人也不反对"金窝银窝不如自己的狗窝"的自在型豪宅。

要做一位有能力提出标准的设计师

有能力的消费者对住宅的要求更多、更严格，品位更高，规划的内容也更细致，因此身为设计师就必须相对地提升自我的生活品位和艺术涵养，才有能力给予你的业主提升生活品质的设计规划，而不是业主提出标准，设计师消化标准。室内设计师除了是精通专业领域的专家外，也要是懂得生活品位的专家，更是人文艺术的专家。唯有如此，他才足以满足现代豪宅的业主的需求，也才有资本来说服有能力消费的业主。

豪宅施工细节

对于豪宅业主而言，房子当然不是只有简单的居住功能而已，还要既展现业主个人能力，又不乏含蓄内敛，以下简略说明本人对于豪宅施工要求的原则。

1.注重内外兼备的工法：每一个步骤关系到后续的品质，所以必须步步严谨；外表质感虽然重要，但内在才是长远的保证。看得到的是细节，看不到的是一切。

2.安全第一的设计要求：住宅即城堡，没有安全其余就一切免谈；法规中的防火标准是最低要求，我自己定出超高的安全标准，唯有如此才对得起业主，我才能安稳入眠。

3.做别人没做到的事情：别人已经在做的是基本的工法，我们做的是意想不到的和深入的一面。例如，电路无交流声、冷气无冷击风，所有家具的尺寸符合业主特点的人体工学，灯光保证具有三种功能（基础照明、工作照明和情境照明），以及对地理环境的考量，同时还要兼顾业主个人爱好，展示其收藏品，展现多元的设计。

4.科技与未来的前瞻设计：云端设备的预期规划整合、监视与监控的稳定方便、视听音响设备的布线要求及用水用电的管道设备维修预留。

5.空气和水的品质监管：要求室内任何一处使用空间的空气每4~6小时更新，新空气必须过滤粉尘。洗衣、洗浴和饮用的水要经过不同的过滤系统。

就个人而言，"好宅"比"豪宅"更重要！

"好宅"定义

以下是个人根据多年从事室内设计的经验，对"好宅"的定位提供的参考。

1.选择好的建筑条件：好地段、好景观、好邻居、好环保和好建商（上述条件属都会型建筑物）。

2.严格选择好的设计团队：有品位、有信誉、有品德、有气度以及能坚持（室内设计是艺术行为，而不是工程）。

3.细选好的建筑材料：好品牌、好口碑、好性能和有利于健康。

4.艺术涵养：独到的艺术鉴赏眼光、脱俗的品位格调、深厚的人文素养，而非泛滥成灾、盲目跟风而沦为流俗。

5.富于未来感的简洁布局：重环保、少维修、减能源以及内涵丰富才能感染业主。

雍容中式
GRACEFUL CHINESE STYLE

010	沉醉艺术空间，蕴养尊荣气度
	AN ARTISTIC SPACE TO CULTIVATE DIGNITY AND GRANDEUR
024	中西合璧，雍容府邸
	A COMBINATION OF EASTERN AND WESTERN CULTURE IN A GRACEFUL MANSION
034	金风玉露一相逢，便胜却人间无数
	A PERFECT COMBINATION LEADS TO A PARADISE
040	阑干倚处月华生
	RESTING BY THE RAILING, BATHED IN THE MOONLIGHT
048	清爽舒适，淡雅中式
	CRISP AND COMFORTABLE, ELEGANT AND CHINESE-STYLE
054	国学精粹，东方美境
	THE ESSENCE OF CHINESE ANCIENT CULTURE AND THE AESTHETICS OF THE EAST
064	传承丽江古韵，顺应山形水势
	INHERITING THE ANCIENT CHARM OF LIJIANG; FOLLOWING THE CONTOURS OF ITS MOUNTAINS AND RIVERS
072	海上四合院
	A QUADRANGLE ON THE SEA
084	艺术空间，融汇古今
	A COMBINATION OF CLASSICISM AND MODERNITY IN AN ARTISTIC SPACE
102	能住的博物馆
	A LIVABLE MUSEUM

浓情东南亚
BRILLIANT SOUTHEAST ASIAN STYLE

120	时尚撞上古典，迸发风格火花
	WHEN FASHION MEETS CLASSICISM
128	大气格局，尽抒写意情怀
	MAGNIFICENT MANSION FOR SPIRIT CULTIVATION
134	自然东方，生活长青
	THE NATURAL EAST, THE EVERGREEN LIFE
142	光影游弋，自在闲庭
	THE DYNAMIC LIGHT, THE LEISURELY COURTYARD

150	深紫幽蓝，浓情美居	
	THE BEAUTIFUL RESIDENCE OF PURPLE AND BLUE	
156	回归自然的东南亚风情	
	RETURN TO THE NATURAL SOUTHEAST ASIAN CHARM	
162	花果飘香，浓情雅居	
	THE ELEGANT RESIDENCE IN MOOD AND FRAGRANCE	
168	木色倾城，闲情逸居	
	WOODEN LEISURELY RESIDENCE	
174	恋上东南亚，回归大自然	
	FALL IN LOVE WITH SOUTHEAST ASIA, RETURN TO NATURE	
182	别样混搭风，闲情东南亚	
	THE TREND OF UNIQUE MIX AND MATCH, THE LEISURE OF SOUTHEAST ASIA	
192	海天别墅，岛上的隐逸假期	
	VILLA EMBRACED BY SKY AND SEA, LEISURELY VACATION ON THE ISLAND	
208	热带花园，坐享印度洋海景	
	ENJOYING VIEWS OF THE INDIAN OCEAN IN A TROPICAL GARDEN	
212	凭栏处，自然风景如画	
	COMMANDING A FINE VIEW OF LANDSCAPE	
218	本土风情，融入时尚设计	
	THE LOCAL ELEMENTS INTEGRATED INTO THE FASHIONABLE DESIGN	
236	传统与现代的无缝对接	
	THE SEAMLESS CONNECTION BETWEEN TRADITION AND MODERNITY	

神秘阿拉伯
MYSTERIOUS ARAB STYLE

246	丝路繁锦，古都神韵	
	THE SILK ROAD, THE ANCIENT CHARM	
252	豪华之巅，尊贵无限	
	THE UTMOST LUXURY, THE EXTREME NOBLENESS	
258	糅合奇幻梦想与不朽传统	
	A MIX OF FANTASY AND MONUMENTAL TRADITION	
298	异域风情，圣洁宁静	
	THE EXOTIC THEME, THE HOLY AND TRANQUIL ATMOSPHERE	
306	瑰丽想象，沙漠绽放	
	THE ROSE BLOSSOMS INTO IMAGINATION IN DESERT	

GRACEFUL CHINESE STYLE

雍容中式

沉醉艺术空间，蕴养尊荣气度
AN ARTISTIC SPACE TO CULTIVATE DIGNITY AND GRANDEUR

◆ 项目名称：中国台湾尊胜白金苑样品屋
◆ 设计公司：动象国际室内装修有限公司
◆ 设计师：谭精忠
◆ 参与设计：詹惠兰、陈敏媛
◆ 面积：346.5 m²
◆ 材料：喷漆、镀钛、钢刷木皮、贝壳马赛克、壁布、钢刷木地板、石材、夹纱玻璃、灰镜、雪花石

◆ Project Name: Paramount Platinum Court Sample Flat in Taiwan, China
◆ Design Company: Trendy International Interior Design
◆ Designer: Tan Jingzhong
◆ Assistant: Zhan Huilan, Chen Minyuan
◆ Area: 346.5 m²
◆ Material: spray paint, titanizing, wood veneer with brush stroke, shell mosaic, wallpaper, wood flooring with brush stroke, stone, yarn-clipped glass, gray mirror, alabaster

本案位处台北市中山区精华地段，为少数拥有便利交通及完善生活设施的新建个案之一，是都市生活中理想的居住环境。低调而洗练的设计语汇、内敛而富于质感的材质与色调搭配，营造出空间的大气感觉，彰显出本案的设计重点：精致生活、家具和艺术。本案是这三者高度融合的展现舞台。

玄关

玄关以整体壁面与线板整合，展开进入样品屋的序曲。而在简练的造型壁板内，另藏有壁橱，兼具收纳衣帽与鞋子的功能，将原有的结构巧妙地隐藏在壁橱中，除满足对收纳空间的高度需求外，更具备了很强的实用性。

客厅、餐厅、厨房

由玄关进入客厅、餐厅区时，即能感受贯穿客厅与餐厅的开放式空间的宽敞与气度。设计师以新的设计表现手法来连贯客厅、餐厅及厨房，搭配运用线板、壁板、斗框等材料演绎新东方语汇，呈现豪宅气势。接待客厅主墙以洞石搭配线条简练的嵌入型的石材壁炉，钢刷木皮质感的壁面造型在线板与石材踢脚板的点缀下，创造出独有的视

觉韵味，也突显出精致的细节处理，铺陈空间的张力与层次感。精致的家具与收藏的艺术品交相辉映，呈现雅致的美学氛围，体现出与艺术装置亲密互动的设计思维。

开放式的餐厅结合中岛厨房的轴线设计，具有空间一贯的流畅性。天花板照明设计应用夹纱玻璃材质，透出柔和的灯光，加上搭配齐全的料理设备，不论是小型聚餐还是大宴，都能让客人舒适地享受美食风味。

主卧室、更衣室、主浴室

主卧室的设计维持一贯舒适的基调，钢刷木皮的主墙面，搭配镀钛与装饰铆钉的点缀，再次呈现出主卧室的精致与典雅。更衣室可收纳行李箱、棉被等物品，抽屉内贴心地设计了绒布格，方便放置首饰、手表、眼

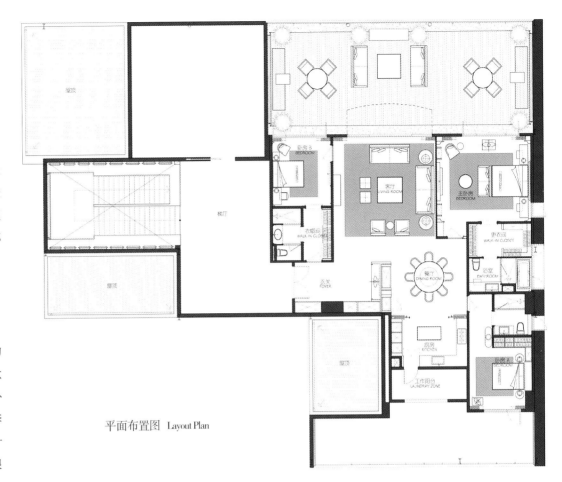

平面布置图 Layout Plan

镜等贵重物品。实用贴心的收纳功能，体现出如精品专柜般的品位，反映出优雅的生活模式，所有物品完整分类，将更衣室空间功能表现得淋漓尽致。

主浴室除了以宙斯石材搭配黑网石镜框衬托高品质的卫浴设备外，更拥有干湿分离的浴厕空间、有梳妆台功能的台面水槽，使人感觉如同置身于高档饭店。洗手台区也加强了坐式化妆台的功能，希望将住家养身的安全观念贯彻于本案之中。主浴室的设计兼顾功能的同时又能保持空间的完整性，为现代社会高度忙碌的人士提供体贴的服务和身心的充分享受。

卧房A、浴室

卧房A温馨、稳重、舒适，材质上运用木皮表现出细致温润的质感，使人迫不及待想沉溺于此舒适睡眠空间，如此细腻设计构成完美

至极的空间。

走入浴室空间映入眼帘即是大面明镜,增加了空间的张力;内置储物的收纳空间,也创造出极富表情的浴室空间。

卧房B、更衣室、浴室

线条干净的空间里,延续框架的设计语汇,巧妙地整合床头主墙面,让卧房的线条层次更加立体。墙面壁纸搭配温暖的橡木钢刷地板,不论昼夜都能有最佳氛围的呈现。

更衣室功能完整且精美,与浴室空间相连接。衣柜则采用功能性的设计方式,希望能在有限的空间中做最有效的设计。镀钛的镜框搭配雪花石壁灯,让浴室空间不仅在视觉上呈现干净氛围,还增添了高贵之感。

Located in the greatest section of Zhongshan District in Taipei, this new project, as one of a few projects which have convenient transportation and well-equipped facilities, is ideal for city living. With the reserved and succinct design language, and a palette of restrained and textured materials and colors, the grand and generous sense of the space is then shaped, which eventually makes a stage for a perfect combination of an elegant life, furniture and art, the key design of this project.

Entrance

The sample flat starts with the entrance in good connection of the whole wall and the line boards. Inside the concise siding of special shape is a built-in closet for clothing and shoes. The original structure is skillfully hidden in the closet to meet the high demand for storage and to offer good practicality.

Living Room, Dining Room and Kitchen

Behind the entrance, the living and the dining rooms firstly strike people with their openness, generosity and spaciousness. With new ideas, the designer creates continuity of the living and the dining spaces and the kitchen, and interprets the new oriental design language with line boards, sidings, and the bucket box frames to display the grandeur and generosity. The main wall in the reception room is equipped with recessed stone fireplace of simple lines, accompanied by travertine. The wall of wood veneer with brush stroke is embellished with line boards and stone skirting board, creating unique visual charm, highlighting the fine details and displaying the spatial tension and layers. The fine furniture and works of art set off each other, showing an elegant aesthetic atmosphere available in a residential space and interactive with art installation.

In addition to the consistent smoothness of the space, the open dining room fits in well with the axis of the kitchen island, and the lighting of the ceiling coated in yarn-clipped glass casts soft light, coupled with a full range of cooking equipment which is available for both a dinner party and a large banquet and allows guests to enjoy the gourmet food.

Master Bedroom, Changing Room and Master Bathroom

The design of the master bedroom is consistent in a comfortable tone, where the main wall of wood veneer with brush stroke decorated with titanizing and decorative rivets represents the exquisite taste and elegance of the master bedroom. The changing room

is also for accommodation of luggage, cotton quilt and other items, where the drawers are equipped with lint grids, appropriate for valuables like jewelry, watches, glasses and others. The practical and considerate function of collecting items with a boutique counter-like taste, reflects an elegant lifestyle. When a complete classification is made, the functions of the changing room are maximized.

Apart from the high quality sanitary equipment in the background composed of the Zeus stone and black mesh stone mirror frame, there is a partition of wet and dry areas and a sink top that can function as a vanity table in the main bathroom, which allows people to feel as if in a high-class restaurant. The sink area has additionally been facilitated with a sitting dressing table, aiming to carry out the concept of home cultivation and security. The design of the main bathroom has taken into account both the functions and the integrity of the space, offering caring services and full enjoyment of body and soul to the hectic people in modern society.

Bedroom A, Bathroom

Warm, stable, and comfortable, bedroom A is of a texture that is warm and delicate with wood veneer, so that the people inside can't wait to indulge in this comfortable sleeping space. Such an exquisite and precise design contributes to a perfect space.

The bathroom strikes people with a large mirror, which increases the tension of the space, and the built-in storage creates a very expressive bathroom space.

Bedroom B, Changing Room and Bathroom

With clean lines, the bedroom continues the design language of framework, skillfully integrating the main wall behind the bedside, so that the layers of the lines look more three-dimensional. The texture of wallpaper sets off the warm oak flooring with brush stroke, bringing out a best atmosphere around the clock.

Fully functional and elegant, the changing room is connected to the bathroom. The wardrobe is approached with functional design, with an aim to design this limited space as the most effective one. The titanizing mirror frame and wall lamp of alabaster make the bathroom look clean and add a touch of nobleness to it.

中西合璧，雍容府邸
A COMBINATION OF EASTERN AND WESTERN CULTURE IN A GRACEFUL MANSION

- 项目名称：中国台湾豪宅
- 设计公司：奥迪国际室内设计
- 设计师：杜康生
- 摄影师：范宸雄
- 面积：500 m²
- 材料：天然石材、银箔、皮革、云石、特殊砖
- Project Name: A Mansion in Taiwan, China
- Design Company: Audi Design
- Designer: Du Kangsheng
- Photographer: Fan Chenxiong
- Area: 500 m²
- Material: stone, silver foil, leather, marble, special brick

排列与组合几乎是所有设计中不可或缺的美学元素。引领新东方风格时尚的顶级空间，除了有行云流水般的动线规划和气势雄浑的布局外，当然也少不了搭配得当的经典家具和各式东方风格的艺术品。考究的细节不仅体现出东方文明在历史长河中沉淀出来的璀璨结晶，而且体现出豪宅主人的不凡品位与美学鉴赏力，同时也是掌握顶尖时尚主导权的关键所在。

奥迪国际室内设计一贯坚持精益求精，且具有丰富的豪宅规划经验，受到许多社会精英的热烈追捧，稳坐知名设计师的名单前列。设计总监杜康生表示：这些豪宅的拥有者，都是清一色的事业成功的企业家，想受到他们的青睐，绝不能只设计出一般意义上的好作品，更重要的是需要高度的定制化与独一无二的艺术性。奥迪国际室内设计通过团队的紧密合作，巧妙地结合每位豪宅主人不同的品位与爱好，以超乎想象的高标准，提炼出类拔萃的感官意象，让无与伦比的尊贵与独特顺理成章地成为业主身份与人生价值的象征。

主人特别喜爱收藏各种艺术作品，尤其是世界闻名的当代顶尖琉璃艺术家王侠军的大型限量版作品，几乎一件不缺。要在不同的空间中妥善珍藏和展示这些尺寸和重量都非常可观的作品，无形中大大增加了整体施工的难度。一进门，几乎全以精选石材打造的独立玄关气势非凡，上方天花板用贵重的银箔覆盖在表面，细腻的光泽有助于延伸视野，显得深邃和气派。动线右侧靠墙处有内嵌式精美展示柜，在中段处搭配一体成型的石材边框，提升了结构的强度，增加了立面层次的造型变化。灯光与后衬镜面反射的光芒，带给人们强烈的视觉冲击。经由别致的造型门拱进入公共空间，设计者在直行动线的尽头特别放置了一个巨型黑色石材展示柜，用来摆设100 kg以上的大型琉璃艺术品，并搭配带有精美的带状造型的天花板，在巧妙地引导视线的同时，又完成对客厅、餐厅和厨房的功能划分。

客厅气象恢弘，超越奢华定义。设计者以精致绝伦的中式风格石材拼花取代传统地毯，并善用简化的古典线条诠释空间立面造型，让线条层次显得低调而富于立体感；运用中西合璧的设计手法营造空间，使其成为凸显精品家具与众多艺术品的绝佳背景。大气的主墙跳出一般电视墙规划的框架，结合用双色石材精工打造的磅礴边框，以厚实沉稳的不动姿态，衬托出中央及两侧展示格柜光芒流转的多层次美感。天花板附设内嵌的大型投影荧幕，搭配多声道环绕立体声顶级音响组合，满足主人待客或独享时对影音设备高标准的要求。

如果说流畅的空间布局是全案的精华，那么遍布全宅、形质俱佳的多款当代名贵装饰与各式艺术品，无疑是活化空间、画龙点睛的灵魂所在，当然也体现出主人卓越的品位和艺术鉴赏力。除了引人注目、独具特色的琉璃艺术品外，屋子里皆是当代顶尖设计名家的经典作品。客厅一张出自美国赫赫有名的J. Robert Scott的手笔的典雅米色皮革沙发和来自意大利Poltrona Frau的暗红马鞍皮革沙发相得益彰，而中央的铜雕方几同样大有来头，是有"美国十大最佳设计师"美誉的Robert Kuo的作品。明亮的餐厅与吧台区也不遑多让，除了气势恢弘的镶银箔圆形穹顶天花造型所对应的意大利知名品牌Cassina的餐桌椅组合外，最引人注目的是一个在餐桌旁靠墙摆放的意大利Promemoria餐柜，其设计深受中国

禅风影响，柜体选用珍贵的木料，纯手工制作，极度耗工费时，细腻的纹理让整体气质深蕴内敛，这是真正行家才懂得收藏的家具艺术极品。

禅风浓郁的和室里，也能欣赏到跨界艺术的精髓。在格栅纸拉门掩映的天光下，一张黑色漆面的和室桌的桌面上，一对金色锦鲤仿佛随时会腾跃于空中，这是日本大师小西起介的不传之作。主卧室内附设的女主人书房，里面有一张浑圆、优雅的书桌，厚实的主体由四只纤细的兽足不可思议地承托着，显示出高超的工艺水平，这是意大利Ceccotti的代表作。不容错过的还有主卧室内豪门专用的Hastens蓝白格纹床组，简洁的外观由多种顶级材质打造，与生俱来的极度舒适无可言喻，它不仅是瑞典皇家指定的品牌，更是全球名门望族的寝室最爱。

主卧室以清雅的纯白为主调，利用横向勾缝线条扩张空间感。床头区洗练的皮革分割造型，穿插迷人的间接光源，透过雅致温暖的实木天花板造型边缘轻轻撒下，为空间抹上一层低调的奢华气息。时尚的床头立面设计与床尾视听功能墙相呼应，两侧则以视线可穿透的镂空木屏作隔屏辅助，利落地界定后方的书房区。除此之外，生活上的实用性也被纳入考虑范围内，超星级主浴空间的精美石材搭配皮革砖，彰显无与伦比的新东方风格居宅的时尚特点。

Arrangement and combination are indispensable aesthetic elements for nearly all designs. Apart from fluid circulation planning and magnificent layout, the well-matched furniture and various works of art of eastern style are indispensable for a top residence which leads the trend of neo-eastern style. As the key to leading fashion, the delicate and precise detailing not only reflects the essence of the eastern culture over a long history, but also reveals the very good taste and the aesthetic sensibility of the owners.

With continuous improvement and innovation, and a wealth of experience in residential planning, Audi Design is remarkably popular among a large group of social elite and has been leading among renowned designers. According to Du Kangsheng, the Design Director, all of the owners of mansions, who are successful entrepreneurs, can be conquered not only by fine works in a general sense, but also by high customization and unique artistic quality. With good teamwork, Audi Design skillfully takes account of varying tastes and preferences of the owners and refines outstanding sensory imagery with high standards beyond people's imagination, so that there is no doubt that the unparallel nobleness and uniqueness of the mansions present the social status and value of the owners.

The host of the mansion is passionate for art collections, in particular the large works of limited edition, which are nearly complete and created by Wang Xiajun, a top world-renowned colored glaze artist. It is a challenge for the designer to create a suitable space where the host can treasure up and display the sizable and heavy works of art in different parts of it, and the execution seems to be harder than before. The separate entrance almost made of specially-selected stones is magnificent. Above the entrance is the ceiling coated in silver foil, whose fine luster extends visions and makes the space deep and imposing. On the right side of the circulation and next to the wall there is an exquisite recessed display cabinet, the middle of which is framed into a piece of stone, which strengthens the structure and changes the shapes of the layers of the elevation. The light and the reflection of the mirror that serves as a background combine to have a great visual impact on people. Behind the special arch is the public area, where the end of direct circulation is equipped with a large black stone cabinet for displaying the large works of colored glaze which are over 100 kg in weight, accompanied by a ceiling with an exquisite stripe-like modeling, skillfully guiding vision and defining the living room, the dining room and the kitchen according to their functions.

The grandeur and generosity of the living room are beyond luxury. On one hand, the designer uses exquisite Chinese-style stone parquet instead of traditional carpets and designs the elevations with simplified classical lines, which are made reserved and three-dimensional; on the other hand, he creates a space with Chinese and western styles, so that the space has become a perfect background of the high-quality furniture and many works of art. Beyond the framework of ordinary TV walls, rimmed by a sizable frame carefully made of stone of two colors, the grand main wall functions as a solid and steady background of the display cabinets in the middle and on both sides of the space, bringing out their brilliant beauty of various gradations. Inside the ceiling there is a large recessed projector with a multi-channel and surround sound top stereo system to meet the enormous demands for audio and video when the host entertains his guests or stays by himself.

If the whole space features its fluid and airy layout, then a variety of contemporary precious furnishings and works of art with good appearance and quality are applied throughout the whole mansion, adding dynamism to the space and highlighting it, so that the owner's remarkable taste and artistic sensibility are therefore reflected. In addition to the eye-catching and unique colored glaze works of art, the classics designed by the contemporary masters can be seen around the mansion, including a classical beige leather sofa by J. Robert Scott, a famous designer from America, which sets off the dark red sofa of saddle leather by Italian Poltrona Frau, and the tea table in the center, which is carved in bronze by Robert Kuo, one of the best ten American designers. The bright dining room and the bar area are as splendid as the above spaces. Apart from the tables and chairs of Cassina, an Italian well-known brand, which correspond to the grand dome ceilings coated in silver foil, the dominant feature of the space is a buffet of Promemoria, an Italian brand, which sits next to the table and the wall and is strongly influenced by a Chinese Zen style. The hand-made buffet is made of rare material. It takes a long time to make the buffet, a piece of treasure exclusively for real collectors, which appears reserved and implicit with its delicate texture.

The room where people can sit on the Tatami is of strong Zen style. It also allows people to appreciate the essence of the combination of eastern and western art. The sunshine is cast through the paper interlayer sandwiched by the grills of the sliding door onto the black lacquered table in the room, on which a pair of golden carps seem to be readily leaping into the air. It is created by Keisake Konishi, a Japanese master. Inside the master bedroom there is an attached study for the hostess, where the solid main part of a round and classical desk is incredibly supported by four delicate feet of an animal, bringing out its excellent craftsmanship. It is the representative work of Ceccotti, an Italian brand. As a privilege of those rich and powerful families, the blue white check bed group of Hastens are too great to pass up. The concise appearance of the bed group is made of various high-quality materials, which provide ineffably extreme comfort. It is not only the designated brand exclusive to Swedish royal families but also the favorite of global notable families.

The master bedroom is dominated by white tone, where horizontal pointing lines expand the sense of space. The succinct leather split-modeling around the area of the headboard is interspersed with indirect light that is cast through the edge of the delicate and warm ceiling made of solid wood, adding a reserved but luxurious touch to the space. The stylish appearance of the head of the bed corresponds to the wall with stereo and video equipment at the end of the bed, while on both sides of the space are the hollowed-out wood ware partitions which allow people to see through them. All of the above clearly define the study in the rear of the mansion. Besides, the practicality is also taken into consideration. The upper-star master bathroom with a palette of stone and leather brick is bound to highlight the unparallel fashionable characteristics of neo-eastern-style residences.

金风玉露一相逢，便胜却人间无数
A PERFECT COMBINATION LEADS TO A PARADISE

- 项目名称：中国台湾中悦知音
- 设计公司：岩舍国际设计事务所
- 主持设计师：林济民
- 执行设计师：吴家桦
- 面积：336 m²
- 材料：欧洲樱桃木、非洲柚木、壁布、茶镜、柚木地板
- Project Name: The Intimate Residence in Taiwan, China
- Design Company: Anta Design
- Chief Designer: Lin Jimin
- Executive Designer: Wu Jiahua
- Area: 336 m²
- Material: European cherry, African teak, wall cloth, tea mirror, teak flooring

本案一进入玄关便以精致的欧洲樱桃实木条镂空墙面造型及金箔饰柜吸引目光，旋即将人带入室内空间，映入眼帘的是与玄关的镂空实木窗花设计相同的拉门，原来这是具有人文氛围的书房空间，局部架高的区域是品茗及赏玩收藏品的专属空间，由此塑造了独立且具私密性的主人书房。

客厅电视主墙面同样以实木镂空造型为主轴，其中有隐藏式的收纳柜，局部以灯箱手法呈现，达到营造不同氛围的效果，主墙面更以特殊壁布为中央端景造型，达到不同的视觉感受。沙发背墙以非洲柚木为底，欧洲樱桃木为分割造型，两者巧妙地结合在一起，如同编织般的效果，让空间活灵活现。

客厅沙发背墙另一侧，以富贵树增添绿意，一来强化视觉端点，二来也让木头温润的色系更具活力，让空间更具生机。

在客厅与餐厅之间，以一端景展示柜及名家书法屏风为区隔，巧妙地界定了空间，活跃了中介空间的属性。餐厅墙面一侧以茶镜喷砂图腾造型，延续全空间的语汇，但不同的材质却呈现出不同的意境；另一侧往厨房方向，茶镜以小方格为喷砂图腾，与对向的窗花造型达到相互呼应的效果，同中求异，更添空间的活跃感。

主卧室床头以全幅式的斜板壁纸搭配香槟色立体造型板，拉宽空间尺度，并更添大气格局。电视墙面两侧同样延续公共空间的镂空窗花造型，达到空间的一致性。床边柜及端景五斗柜以方格造型为饰

面，活化空间属性之余，也呈现家具的不同质感。另一侧的古董收藏柜更是体现主人欣赏品位的好地方。

次主卧室巧妙地以造型框隔开两个空间，分别是睡眠区及书房衣柜区。虽然不是极大的空间，但是善用空间，也能营造出不同的生活趣味，并划分出具有不同使用功能的区域。

客房以基础的使用功能为主轴，强化空间的实用性，以香槟色系为空间主色，让小空间也具有大气的氛围。

The space firstly strikes us with the exquisite hollowed-out wall of European cherry bars and cabinets of gold foil surface at the entrance. In the interior, the sliding door of window grilles continues the design of the hollowed-out solid wooden window grilles at the entrance. The space is a study with cultural atmosphere, and the partially elevated area is specially for tea drinking and collection admiring, which forms the private and independent master study.

The TV wall in the living room takes the hollowed-out solid wooden modeling as its axis. Inside it the hidden storage cabinets are partially presented in the approaches of light boxes to create an effect of various atmospheres. The main wall brings out different visual experiences with a special wall cloth to make the central end view. Meanwhile, the wall behind the sofa is covered by African teak and is split by European cherry. Such a clever combination looks as if it is woven, and therefore the space comes alive.

The other side of the sofa wall of the living room is decorated with Robinia Idaho to strengthen the visual endpoint with the greenery and to make the wooden warm hue more active and the space full of vigor and vitality.

Between the living and the dining rooms stand a display cabinet that serves as the end view and a screen inlaid with calligraphy by famous artists. The space is therefore defined and the medium space has become dynamic. The tea mirror with sandblasted totem on one side of the wall in the living room continues the design of the whole space, but the different material conveys a different artistic concept, while the other side towards the kitchen is decorated with tea mirror of lattice still with sandblasted totem, echoing with the opposite window grilles. Such a treatment retains different features in the cohesive design and makes the space more active.

The bedside of the master bedroom is coated in inclined board fabric of full size with three-

平面布置图 Layout Plan

dimensional modeling of champagne panels to visually enlarge the space. Both sides of the TV wall continue the style of the hollowed-out grille in the public space to achieve a spatial consistency. The Bedside drawers and the cabinet that actually serves as end view are of lattice for the activation of spatial attributes while bringing out different textures of furniture. The cabinet for antique collection is a good place that embodies the taste of the owner's appreciation.

The secondary master bedroom is framed into a sleeping area and a study/wardrobe area. Though not big, the space with flexible functions creates a different life and is divided into different functional areas.

The guest room takes the basic function as the axis and strengthens the use of space. It is mainly in champagne hue to give the small space an atmosphere of magnificence and grandeur.

阑干倚处月华生
RESTING BY THE RAILING, BATHED IN THE MOONLIGHT

- 项目名称：生茂养生园样板房
- 设计公司：于其琛空间设计、大珏国际
- 设计师：于其琛
- Project Name: The Show Flat for Shengmao Life-nourishing Garden
- Design Company: Eison International Design / Deco International Design
- Designer: Eison

生活中有美，自然中也有美，本案的设计就是在生活和自然中取得平衡，寻找美的感觉。设计师在业界积累了丰富的别墅设计经验，专注于别墅设计的同时，以现代的手法表现了中国历代经典的红色系，从而营造一种富有人文情趣的美与韵致，同时又传递当代生活的美学，并且凝固中式的记忆与文化。

生活中的美学

该案主要以中式传统风格与当代生活相结合,将美好的文化雅韵实践到当代生活中。客厅的天花板装饰与屏风的线条遥相呼应,沙发的色调质朴、典雅,地毯的怀旧气息表达了清雅含蓄、端庄丰华的东方精神境界,很自然地使得古典、庄重、大气、沉稳集于一个空间展示出来。而餐厅室内外景致和谐相融,室内木质家具的红与室外草木的绿相得益彰,吊灯与餐椅的颜色又搭配协调。为表现中式风格,所选用的家具材质强调最大限度地贴近自然,像空间中珍贵的陈年榆木、厚重的檀木、花梨,这些既与传统相关,又与人亲和的天然素材,在自然的气味与真实质感中,体现出令人感动的朴素之美。尤其是这些材料的清新气息和经过时间沉淀后的质感,蕴含自然的生活和单纯的呼吸,让人们真实地体会到生活与时间的意义。

浓郁中式笔韵

"柳外轻雷池上雨,雨声滴碎荷声。"把欧阳修的《临江仙》刻在居室的背景墙上作装饰,配合室外的竹景,可谓别有洞天。具有浓郁中式古典风格的瓷器安置于居室里,让愉悦精神的装饰成为家中经典的点睛之笔。加上青花瓷器的摆设,给人清雅的感觉,使得房间拥有一种别样的怀旧之情。在主卧室,设计师利用金色的纱幔增加了其柔美气息,一改中式的沉稳、封闭之感。天花板的处理也是采用自然、简单的线条装饰手法,配合红色的中式风格的背景墙,呈现出一种尊贵的生活品位。

除了居室的中式风,浴室里的中国风也随处可见,红色的木质装饰、浴柜与直线切割的大理石相结合,目的是创造一个宁静而舒适的生活环境。设计师就是采用这种中式风的手笔,将现代元素和传统元素结合在一起,以现代人的审美需求来打造富有传统韵味的室内设计,让传统艺术在现代生活中得到合适的体现,为我们呈现了一个贯穿中西、融汇古今、富有大宅气派的品位家居生活空间。

Life cultivates beauty, so does nature. This project is aimed at attaining the beauty in the balance between life and nature. Committed to villa design, Mr. Yu, a designer, with a wealth of experience in villa design, turns to modern techniques to bring out Chinese classic red hue which has been inherited since ancient times, creating a kind of beauty and charm abundant in cultural taste while conveying the aesthetics of contemporary life, and solidifying Chinese memory and culture.

The Aesthetics of Life

The project, a combination of traditional Chinese style with contemporary life, is a practice of culture and elegance in modern life. The decoration of the ceiling of the living room corresponds to the screen lines; the sofas are in simple and elegant hue; the nostalgic sense of the carpet is an image of the Chinese spiritual realm, which conveys the qualities of elegance, implication, graciousness and generosity, leading to a combination of classicism, solemnity, grandeur, and solidity shown in a space. The views inside the dining room keep in harmony with the landscape outside. The red wooden furniture indoor matches the green vegetation outdoor, while the chandelier and the chair match in colours. The chosen materials for the Chinese-style furniture are actually close to nature as much as possible, such as the old elm, the ebony, and the rosewood. With relation to tradition and accessibility to people, all of the above natural materials reflect the impressive beauty of simplicity out of their natural smell and their real textures. In particular, the fresh smell and the weathered texture of the materials combine to bring out the sense of natural life and pure breath and to allow people to truly understand the meaning of life and time.

The Rhyme Rich in Chinese Style

The backdrop in the bedroom engraved with a poem by Ouyang Xiu, a poet in Song Dynasty, accomplishes a good scene with the views of bamboos outdoors. The spirit-pleasant decoration of porcelain with Chinese ancient style becomes a finishing touch, especially coupled with the Chinese blue and white ones, which add a distinctive sense of nostalgia. The master bedroom is applied with gold mantle that changes the steady and enclosed sense of Chinese style. Decorated in approach of natural and simple lines, and accompanied by a red Chinese-style backdrop wall, the ceiling presents a distinguished quality of life.

The Chinese style applied in the living room continues in the bathroom. The red wooden decorations, the bath cabinet and the straight-cut marble combine to create a quiet and comfortable living environment. In these Chinese-style approaches, the designer combines modern and traditional elements to make an interior design rich in traditional flavor to meet the modern aesthetic needs, so that traditional art has found their proper way in the modern life and a delicate lifestyle is therefore presented to us with combination of Chinese and western culture, integration of classicism and modernity, and grandeur of mansions.

清爽舒适，淡雅中式
CRISP AND COMFORTABLE, ELEGANT AND CHINESE-STYLE

◆ 项目名称：东莞万科·棠樾澜山居样板房
◆ 设计公司：深圳市昊泽空间设计有限公司
◆ 设计师：韩松
◆ 面积：100 m²
◆ 材料：戈壁沙漠石材、橡木面板
◆ Project Name: Sample Flat of Vanke Tangyue Lanshan, Dongguan
◆ Design Company: Shenzhen Haoze Space Design Co., Ltd.
◆ Designer: Han Song
◆ Area: 100 m²
◆ Material: stone, oak panel

该案地理位置远离繁华市区，周边环境风景如画，由此定位为周末度假型公寓。原来狭长纵深的空间里，餐厅、客厅、卧室、阳台一气贯通，虽然视线上比较通透而且利于空气流通，但从居住方面来说，有些遮掩、转折和回旋的空间才更佳。设计师特别在卧室外加了一道可推拉折叠的门，打开时光线通透，关上时，卧房自成一隅。

万科·棠樾澜山居周边环境优越，林木茂盛，山清水秀。为了表现一种世外桃源的慢生活，设计师使用了简约化的中式风格来营造一种格外轻松、休闲的氛围。古典的中式设计难免感觉过于厚重，对空间尺度的要求也较高。设计师化繁为简，用米白色来减轻中式的厚重，并且选用简化的中式坐椅、茶几来装点空间，使得文化的意蕴还在，形态却变

得清新明快。

空间的多处细节，也以中式格调的软装搭配出舒适、愉悦的感觉。除了浮雕花纹的麻质墙纸、粗糙的砖石地面、沉稳的柚木木质外，还应用了青花瓷、鸟笼式吊灯、竹制的餐桌椅、天然材质打造的沙发茶几……清新淡雅的设计，让整个空间范围弥漫着些书卷气，稍显慵懒却处处散发着闲散的气息。从卧室角落、客厅到餐厅餐桌，几盆绿意盎然的盆栽，宁静的草木香味，让人能感受自由和遐想联翩，同时展现出别样的现代中式风情……

Its location far away from downtown and its picturesque surroundings make this project an ideal resort apartment for weekends. The original space is long, narrow and deep. The dining room, the living room, the bedroom, and the balcony are naturally cracked open. Despite the visual transparency of the whole space and good ventilation, covers, twists and turns would always make a better design. That's why the sliding door which can be folded stands in front of the bedroom. The light is introduced when it is opened, while the bedroom stays in isolation when it is closed.

This project is ideally surrounded by a lush forest, clear water and green hills. In order to show a paradise-like leisurely life style, the designer creates a particularly relaxing and casual atmosphere with simple Chinese style. And the excessive stiffness out of the classical Chinese design that requires a sizable space is softened with the aid of an off-white hue, with simplified Chinese-style chairs, and a tea table to become fresh, bright and crisp in appearance while maintaining the cultural sense.

A match of many details of the space brings out a comfortable and pleasant feeling, thanks to the interior decoration of Chinese style. In addition to the embossed linen wallpaper, rough brick flooring, and calm teak wood, the elements of blue and white porcelain, bird-cage-like chandelier, bamboo tables and chairs, sofas and tea tables of natural materials make the space filled with an cultural, leisurely and casual atmosphere. From the bedroom corner, the living room to the tables in the dining room stand pots of green plants, giving off a calm and fragrant perfume, which allows people to feel free and enjoy reveries, and shows a unique modern Chinese style.

国学精粹，东方美境
THE ESSENCE OF CHINESE ANCIENT CULTURE AND THE AESTHETICS OF THE EAST

- 项目名称：深圳万科·棠樾会所
- 设计公司：深圳市昊泽空间设计有限公司
- 设计师：韩松
- 面积：3 000 m²
- 材料：木纹灰云石、黑木纹板、柚木、日本纸
- Project Name: Shenzhen Vanke Tangyue Chamber
- Design Company: Shenzhen Haoze Space Design Co., Ltd.
- Designer: Han Song
- Area: 3,000 m²
- Material: gray wood grain marble, black wood grain panel, teak, Japanese paper

万科·棠樾以现代东方美学为导向，以呼应亚洲东方一体化和中国文化的新变化。万科以中式的建筑符号、景观和文化，让代表国学精髓的东方美学意境在建筑细节和空间里找到依靠，实现东方人居的最高境界。其宣传口号是"在东方，生活当以境界甄别"。

万科·棠樾从形式到内涵上均体现了中国文化的精髓，在博采众长的过程中学会扬弃，寻找中国传统居住文化精髓的同时，又吸纳现代生活的时尚元素，实现了传统中式的"神似"与现代人居的"意扬"：带有传统装饰符号但又必须经过现代高科技加工才能达成的"条窗"、以高档的石材面砖组成的青灰色的优雅沉静的高墙和外立面、私家庭院中临水而建的"轩"和庭院外借鉴各种传统园林写意手法营造出的社区环境，创造出古朴、典雅、吉祥、宁静、封闭的居住氛围。室内空间的设计在充分考虑与庭院空间的交流，以及在与地下层、一层、二层垂直空间之间交流的基础上，融入现代生活的特点，符合现代生活的空间肌理，也给整个室内空间带来了通透、亮堂、大气的感觉。

会所的设计与万科·棠樾的定位是相辅相成的。从一开始，本案就以体现中国传统文化，用现代手法进行传神演绎为基调，内外一体，气脉相承。整个空间开阔大气，动线流畅，细节呼应，一气呵成。

万科·棠樾共有三大会所，包括销售会所、运动会所及休闲会所。

运动会所入口尺度开阔，布局疏朗，落落大方。米黄色的石材铺就的墙身带给空间丝丝暖意。入口大门古典纹样的推拉门扇、条案、罗汉床及古色古香的接待台，共同营造出典雅大气的东方格调。

销售会所的接待区以泰式度假酒店式的天花吊顶营造出质朴天然的气氛。接待空间以长桌厚板的中式茶座、绣墩、圈椅、木质布艺沙发做铺陈，让人感到轻松而愉悦。贵宾签约区宽敞的空间让人豁然开朗：几根梁柱、几面透明玻璃墙，再就是三四面可以折叠的镂空屏风式门窗，几组沙发，就构成了一个如此宽敞明亮的空间。室内没有琳琅满目的装

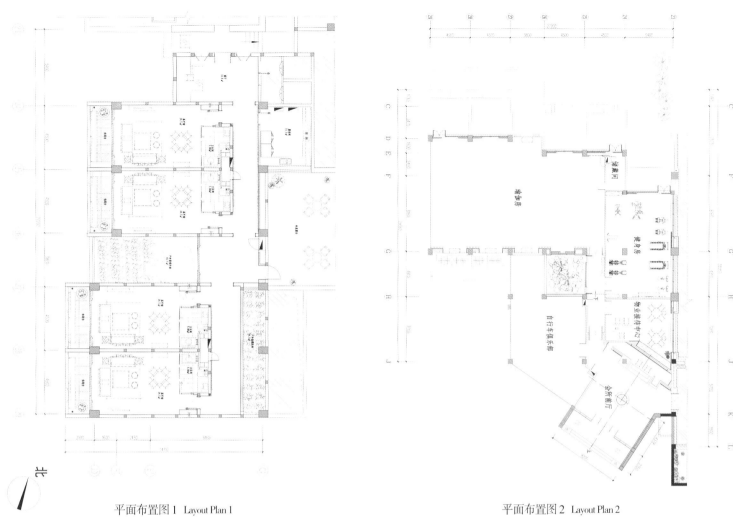

平面布置图 1 Layout Plan 1   平面布置图 2 Layout Plan 2

饰，更没有繁复的线条刻画，但却显得更简约、大气。透过落地大窗，可以纵览窗外绿意盎然、花木茂盛的美丽景致。

休闲会所按会所功能划分为多个区，包括KTV、棋牌室等。KTV接待厅的设计特有娱乐场所的感觉，沙发的深灰和方桌的深黄，颜色的搭配恰到好处的低调，加上具有中国韵味的斜边天花刚好倒映在透亮的地板上，虽是四平八稳的空间，却不失活跃元素点缀其中，符合娱乐空间所呈现的欢快感。

棋牌室的设计既有中国古典韵味，又不失现代轻松氛围。木梁架构，是传统建筑形态的现代演绎。深色的木质地板、镂空推拉门和木质棋牌桌，以自然材质之美，达到了提升整体空间内在品质的目的。鸟笼式吊灯营造出休闲自在的意味，墙壁上数排青花瓷盘的有序排列，增加了空间的趣味性。

针对不同的会所空间，设计师以不同的手法进行演绎，却在总体调性下，保持风格的一致性，让传统的生活情境，进入当代的空间语境中。传统文化不只是拿来观摩或者装饰，而是实实在在地与现代生活融为一体，与时俱进，成为生活本身，这就是空间设计最大的价值所在。

This project is oriented towards modern eastern aesthetics to be consistent with the integration of Asia, and the new changes in Chinese culture. The application of Chinese architectural symbols, landscape and culture allows the artistic concepts of the oriental aesthetics, which present the essence of Chinese ancient culture, to be expressed in architectural details and spaces, thereby achieving the highest level of eastern human settlement. Its slogan is that in the East, life is distinguished in terms of its level.

This project embodies the essence of Chinese culture both in forms and contents and learns widely from others' strong points, which have been promoted while the obsolete is ripped out. It also gets the essence of Chinese traditional habitation culture while absorbing the fashionable elements of modern life, achieving the effect of the resemblance in spirit in traditional Chinese style and the expression of mood in modern residence: the striped–like window with traditional decoration symbols, but made with a high–tech approach, the cinereous, calm and graceful wall and facade that are coated in high–end stone tiles, the pavilion by the water in the private courtyard and the community environment that is accomplished in the freehand method from varieties of traditional gardens outside the courtyard. All of the above combine to create a living atmosphere that is ancient, pristine, elegant, auspicious, tranquil and enclosed. Considering the communication between the courtyard and the interior space, and that between the underground, the first floor, and the second floor vertically, with elements of modern life integrated into it, the interior design fits into the texture of modern space and brings forward a sense of visual transparency, brightness and grandness to the space.

The design of the project supplements the position of the estate. The project is based on the embodiment of the traditional Chinese culture and the modern techniques for expressive interpretation. The internal and the external appear as a whole and cohesive. The whole space is grand and sizable, with fluent circulation and consistent details in a perfect unity.

The chamber consists of three sections: sales, sports and leisure sections.

The entrance of the sports section is sizable with magnificent and excellent layout. The beige stone

wall gives warmth to the space. The sliding gate of classical patterns, the long, narrow table, the arhat bed, and the antique reception desk combine to create an elegant and grand oriental style.

In the reception area of the sales section the Thai resort suspended ceiling creates a pristine and natural atmosphere. The reception area is relaxing and enjoyable, with a long table and a thick plate, a garden stool, a round-backed armchair, and a wooden fabric sofa. And the VIP signing area is really refreshing, where a few beams, several transparent glass walls, foldable hollowed-screen windows, and sofas make a bright and broad space. Without any dazzling interior decoration or complicated lines, the interior space appears simpler and grander. The French windows introduce the greenery and the lush into the inner space.

As for the leisure area, it's divided into several functional areas, including a KTV reception hall, a chess room and so on. The design of the KTV reception hall brings out an entertaining sense. The dark gray sofa and the deep yellow square table implicitly make a good match. Besides, the reflection of the Chinese-style slanting ceiling on the bright flooring lights up the space. This steady space is decorated with dynamic elements, fitting into the exhilarating sense in the entertainment space.

The chess room is filled with a strong Chinese classical flavor while retaining a modern and relaxing ambience. The wooden beams are the modern interpretation of traditional architectural forms. The beauty of natural materials applied in the dark wooden flooring, the hollowed-out sliding doors and the wooden chess tables upgrades the internal quality of the whole space. The cage-like droplight brings out a relaxing and cozy mood, while the rows of blue and white porcelain plate on the wall add fun to the space.

Different spaces here involve various methods, but the style is of consistency. The traditional life is implanted into the contemporary space. Traditional culture is, actually, not only used for observation or decoration, but should be integrated into the modern life to keep pace with the times, and to become life itself, which is the most valuable in the design of the space.

传承丽江古韵，顺应山形水势
INHERITING THE ANCIENT CHARM OF LIJIANG; FOLLOWING THE CONTOURS OF ITS MOUNTAINS AND RIVERS

- 项目名称：丽江和府皇冠假日酒店
- 设计公司：云南省设计院
- 设计师：陈荔晓、陈俊、赵汉鼎、苏宏波、程斌
- 用地面积：52 900 m²
- 建筑面积：49 300 m²
- Project Name: Crown Plaza in Lijiang Hefu
- Design Company: Yunnan Design Institute
- Designer: Chen Lixiao, Chen Jun, Zhao Handing, Su Hongbo, Cheng Bin
- Site Size: 52, 900 m²
- Building Size: 49, 300 m²

丽江和府皇冠假日酒店位于丽江古城与现代城市的衔接区。传统街区的形成是一个历史信息不断叠加的过程，是一种自然生长的力量。因此设计师在该酒店空间布局中力求延伸古城街道肌理，配合古城的地形和水势，一脉相承。客房区南北纵向依水势的曲直变化形成5条主要街道，东西横向穿插次要街道和巷道，顺应山水地势将两种客房单元自由组合，形成了自然有序的街巷空间。这种随意的空间连接方式，避免过度设计带来的呆板和空间紧张感，使空间的变化充满了不确定性，人们在体验中带着期待和惊喜，如同体验古城街巷的自然变化。

两种居住单元设计根据酒店的具体功能要求，将传统民居形式提炼、重构，形成了两种固定单元体，即"L"形拐角单元和联排错位单元。标准的单元体自由生长和交错互动，形成客房区自然有序的空间特性。设计师在单元体的空间中融入庭院、门廊、街区等转换空间，使单元体空间私密与开放适度交替，由此一来，人们可以在安静、放松的状态中体验特有的人文与自然气息。

基地南端的设计有所不同，庞大的公共服务区采用传统庭院的组合方式，化整为零的同时又密切联系，通过多个竖向交通体系把半地下复杂的供应功能和地面上的大堂、餐厅、茶室联系，形成有完备后勤供给的幽静庭院活动区。在这里，人们能体验到现代生活、现代功能空间与传统元素、传统庭院空间的碰撞与交融。

从酒店总体布局到单元的空间设计，均考虑以借景、对景、引导等多种体验方式与雪山景观相呼应。从酒店入口穿过大堂一直到景观露台，

雪山是整个主要空间的视觉轴线，部分客房推开窗后可以远眺雪山，人们在休憩、静卧的时候可以安静地欣赏雪山远景。延续古城的水系，在区域里有独立的水系由北流向南，自然发散在街巷、院落间。有的水流隔断入口，需要越过小桥才能进入客房；有的穿过后院，撞击鹅卵石发出汩汩的声响；有的在街巷转角处形成水榭。流水与建筑的交融，在不经意间传递了古城小桥流水、诗意栖居的意蕴，度假酒店的品质也因为雪山和流水这两个地域特征而提升。

丽江和府皇冠假日酒店不是张扬的地标，而是在古城肌理中延续和生长出来的建筑群体，整个项目与古城没有分界，而是有机地融为一体。建筑群体与古城的人文环境、自然环境产生有机的对话，形成具有地域特征的空间形态，使人们在体验建筑的过程中，感受到地域特征，在心理上形成认同感与归属感。

The Crown Plaza in Lijiang Hefu is located at a threshold between the ancient section and the modern part of Lijiang city. The development of the traditional neighborhood is actually accumulation of historical information and a process that goes spontaneously. Therefore the designers aimed at extending the routes, the layout and the directions of the old streets in the layout of the hotel, and following the water routines and the topography of the ancient city, which makes continuity of the spaces of the hotel and its surrounding context. Following the changes of the water routines, the guest room area is scattered along five main streets, extending from north to south with secondary streets and lanes crossing between, which combine the units of guest rooms of two types freely along the contours of the mountains and the routes of the rivers, so that the natural and orderly space of street is achieved. With such a free spatial connection the space could avoid the stiffness and tension which come out of over design, become uncertain, and allow people to experience with expectations and surprises, as if they were experiencing the natural variability of the old streets.

Two kinds of dwelling units, L-shaped unit and dislocated townhouse unit, are refined from traditional houses, and then reconstructed, but certainly well meet specific hotel functions. The free development of the standard units and the interaction between them combine to contribute to the features of the space of the hotel, so that the whole guest room area is remarkably natural and orderly. Among the unit blocks are the transition spaces, such as the courtyards, the porches, and the blocks, so that the whole unit block are adequately either private or open, where people can be immersed into a unique cultural and natural atmosphere in tranquility and peace.

The southern end of the site is designed differently from the other parts of the hotel. The large public service area is composed of small parts that are connected to each other closely, mimicking the composition of traditional courtyards. The semi-underground supple functions are accessed by the lobby, the dining room, and the tea room on the

ground through the multiple vertical transportation systems, making a complete logistics but a secluded courtyard. Here is indeed a collision and combination between modernity and tradition.

From the overall layout to the interior design of units, the correspondence with the snow-capped mountains is taken into account, through a variety of ways, such as view borrowing, view in opposite place, and eyesight guiding. The snow-capped mountains function as the main visual axis that starts from the entrance, through the lobby and finally to the terrace. Some rooms are open to the views of the distant snow-capped mountains when the windows are opened, and people can enjoy the views in tranquility when sitting or lying. As a continuation of the water system of the ancient city, the separate water system flowing from north to south wanders along streets and through courtyards. Some of the guest rooms are accessed by bridges because of the water in front of them; the pebbles are sounding due to the water flowing through the backyards;

some waterside pavilions are made where the water turns around in the streets. The combination of flowing water and the architecture forms a poetic picture made by composition of the ancient city, water and bridges. The quality of the resort hotel is upgraded by the two topography features, the snow-capped mountain and the water.

The Crown Plaza in Lijiang Hefu appears as a complex, an extension and a product of the ancient city rather than an eye-catching landmark. There is no boundary between the project and the ancient city and the plaza is organically merged into the context of the city. The complex opens up a dialogue with the cultural and the natural environments, making a spatial form with geographical features, so that people can feel the topography and get a psychological identity and a sense of belonging while they experience the architecture.

海上四合院
A QUADRANGLE ON THE SEA

◆ 项目名称：海南香水湾一号
◆ 设计公司：HSD水平线空间设计
◆ 设计师：琚宾
◆ 面积：94 000 m²
◆ 材料：火山岩、桃花芯木、涂料、草编壁纸

◆ Project Name: The No. 1 Quadrangle on the Perfume Bay, Hainan
◆ Design Company: HSD Horizontal Space Design
◆ Designer: Ju Bin
◆ Area: 94,000 m²
◆ Material: volcanic rock, mahogany, paint, straw wallpaper

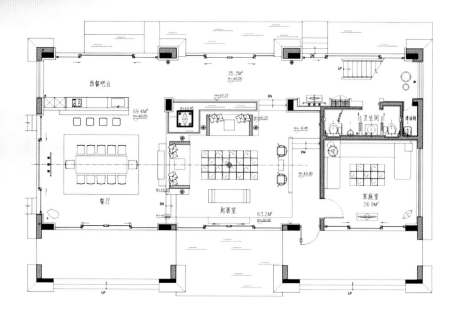

平面布置图 1 Layout Plan 1

香水湾一号是对亲海式度假酒店的研究性项目。建筑的亲海性、对自然的尊重、对室内设计原材料的探索以及科技与智能化的融入，都是设计师研究的课题。

商周时期的建筑元素与现代室内设计相结合，将同时体现建筑的传统性与室内设计的现代性。设计师用"海上四合院"来定义香水湾一号，让建筑敞开怀抱融入自然。设计师提取并保留了中国商周时期的建筑语言，同时重新解读商周青铜器与商周服饰的设计语言，将这些元素运用到室内及家具设计中。材料采用古朴自然的木材和石材，整个建筑与周边环境和谐相容。在这里，左手大海，右手群山，大自然浓缩在一掌之间。

The No. 1 Quadrangle on the Perfume Bay is a research project on seafront resort hotels, including the proximity to the sea, the respect to nature, the exploration of materials of interior design, scientific methods, technologies and artificial intelligence.

The integration of the architectural elements of Shang and Zhou Dynasties and modern interior design reflects the tradition of architecture and modernity of interior design. The space positioned as a quadrangle with its location on the sea is open to nature while architectural language of historical periods is extracted, and then retained, with bronze ware and trappings of Shang and Zhou Dynasties re-interpreted and finally employed into the interior and furniture design. Thanks to the pristine and natural wood and stone, the entire hotel keeps in harmony with the surroundings, which enables people to feel as if merged into nature.

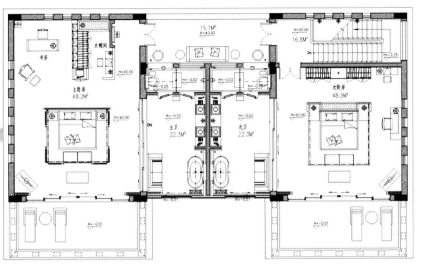

平面布置图2 Layout Plan 2

艺术空间，融汇古今
A COMBINATION OF CLASSICISM AND MODERNITY IN AN ARTISTIC SPACE

◆ 项目名称：四川邛崃文君酒厂
◆ 设计公司：MAP建筑规划公司
◆ Project Name: Wen Jun Distillery, Qionglai, Sichuan Province
◆ Design Company: MAP Architect and Planning Ltd.

设计师的初衷是在地基中央建造一个新的管理辖区，集办公和来访接待为一体。办公室的结构简单，设计优雅，由一楼的"V"形柱体支撑两层"流动的"办公空间。中间设有中庭，覆以椭圆形的大天窗。小型会议室和休闲吧分立一楼两端。休闲吧向木制平台敞开，周边建筑和管理辖区直入眼帘。管理辖区以大型水景为特色，风景秀美。一楼容纳了一个小的演讲厅和酒吧，在木制平台上可以俯瞰下面的景色和邻近的接待中心。一楼门外的水池上铺设了互相交叠的四方形石阶。

来访接待中心按传统酒厂的建筑风格设计，有传统的中国灰色瓦

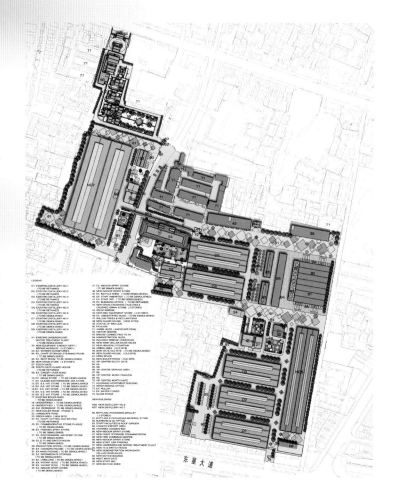

二期工程平面图　Phase 2 Master Plan

顶，以及带装饰的山墙封檐板和支架。墙体由当地的灰砖制作。因为窑的温度有差异，故砖也呈现出各种各样美丽的颜色。二楼VIP品酒室，直通宽大的阳台。从这里出发，带遮蔽的回廊通向酒厂的观赏画廊，宾客在此可以观赏到酿酒的制作工艺，一饱眼福的同时，还能呼吸到酒的醇香。设计师还以传统风格为整个建筑复合体设计了一座新的锅炉房，营造出一种像画廊般的艺术氛围——艺术的锅炉，内有高效、节能的燃气机组。

地基的北部是一个单独的建筑复合体，有独立的出入口，以围护起来的管理区域的形式出现，加上门楼、两栋大型来宾别墅、一栋总统别墅和VIP接待中心，其间各功能设施如休息室、酒吧、餐厅和音乐馆应有尽有。所有大楼均为传统川式建筑，使用灰色砖瓦，而花园围墙则以当地红砂岩装饰表面。水流沿着别墅外墙汩汩而流，衬以每栋别墅入口旁边小瀑布的哗哗作响，感召着远道而来的尊贵客人。如今，设计师接受了委托，根据总体规划设计二期工程，它将包括装瓶和打包的车间，一个物流中心和新的公共观赏廊道。来宾站在廊道中，观赏着标志性的大陶瓮，可以体会到其中丰富的文化内涵。

接待中心剖面图 1　Visitor Center Layout Plan 1

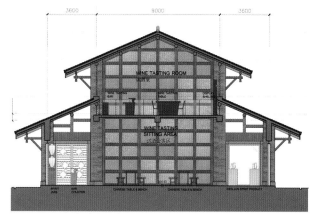

接待中心剖面图 2　Visitor Center Layout Plan 2

One of our first proposals was to establish a new administrative precinct at the centre of the site, accommodating a new office building and a visitor centre. The office was designed as a simple, elegant structure, with V-shaped columns at ground-floor level supporting two "floating" office floors. These are pierced by a central atrium topped with a large oval skylight. The ground-floor accommodates a small conference room on one end and a lounge bar on the other, which opens up on to a timber deck overlooking the precinct and the neighboring visitor centre. The precinct itself is beautifully landscaped around a large water feature. The ground-floor holds a small lecture hall and a bar, serving as an area overlooking the views on a wooden deck and a visitor center nearby. There are also overlapping diamond-shaped stepping stones on the pool outside the ground-floor.

The visitor centre was designed in the style of a traditional distillery building, with a traditional Chinese grey-tiled roof and decorated timber bargeboards and brackets. The walls have been built with the local grey bricks which, because of temperature variations in the kiln, have beautiful color variations. The first floor accommodates a VIP tasting room that opens on to a large terrace. From here, a covered walkway leads to a viewing gallery inside the distillery building, allowing guests to take in the view and aroma of a working distillery. The designers also designed a new boiler house for the whole complex, again in the traditional style, creating an art gallery-like environment for the very latest state-of-the art boilers. There are energy efficient, gas-fired units inside.

The northern arm of the site is given over to an independent architecture complex, which has its own road access; this again takes the form of a walled precinct and incorporates a gatehouse, two large guest villas, a presidential villa and a VIP visitor centre, which houses a lounge, bar, dining room and music pavilion. All the buildings are in the traditional Sichuan style, with grey bricks and tiles, and the garden walls are finished in the local red sandstone. Watercourses flank the external walls of the villas, with a small waterfall next to the entrance of each villa so that visitors are greeted by the sound of flowing water. And now the designers have been commissioned to design plase 2 according to the masterplan. It will include a bottling and packaging plant, a logistics centre and a new public viewing gallery, from which visitors will be able to admire the iconic earthenware vats in which the spirit is stored and aged.

能住的博物馆
A LIVABLE MUSEUM

◆ 项目名称：健一公馆
◆ 设计师：康健一
◆ Project Name: Jianyi Mansion
◆ Designer: Kang Jianyi

都说字如其人，或者画如其人，如果要了解健一公馆，也得从认识它的主人康健一先生开始。康先生热爱中国传统文化，同时也是一位收藏家，在他看来，从一件旧物的细节中感悟历史、触摸历史，是人生最大的乐趣和享受。谈到艺术品收藏方向，做过拍卖的康健一也有自己鲜明的观点，他认为中国油画就像中药行业，目前炒作嫌疑非常大，对此，他并不十分热衷。相对来说，中国水墨画对他的吸引力要大些。在他看来，中国水墨画，从落笔为定到慢慢渗透，产生极其微妙丰富的笔墨变化，运用墨色之变化，强调神韵，追求舍形而悦影、含质而趋灵的艺术境界。自从师承齐派（齐白石）书画之后，他对清末民初的作品更为青睐。自己创作的同时还收藏了大量齐白石的作品。

从这里也可以看出康先生要建造一所精品高端的中式会馆的根源。在他看来，做中式风格的精品酒店，并不是将含有中国元素的物件简单

堆砌，而是在点滴间释放中国文化博大精深的内涵。

循着这条线索，再来欣赏健一公馆，才能发现它的妙趣与精华。

健一公馆坐落于红领巾公园西园，正对5 000 m²室外草坪，其开阔的视野，令人心旷神怡，在寸土寸金的北京，这也是一片不可复制的奢侈之地。所以奔驰、宝马、伯爵表等世界一线品牌都选择与健一公馆合作，在这里举行大型发布会。

健一公馆以中式建筑为设计主体，外观素雅静美，有一种民国建筑的沉稳范儿。公馆的内部设计却是别有一番天地。康先生从故宫中获取灵感，运用到健一公馆的设计当中，从入口大堂、中堂、走廊等公共空间，一直到客房，中式文化中的庄重、静雅、对称、秩序等文化底蕴都涵盖其中。馆内每个角落都是艺术品，处处散发着来自内在的优雅内敛的气质，每一处细节都可以看做中国文化的最好表现。在大堂可以欣赏由木质梁柱架构的天井的庄重气派；行走在木质长廊下，两侧的石雕、砖雕都让人感慨于中华手工艺术的精美；公共空间的木雕艺术和客房的木雕装饰体现出来的繁复的雕刻、精湛的技艺，都令人惊叹。仅有这些还不够，别忘了康先生还是一位收藏家。书画艺术、文化精品等收藏遍布公馆各个角落。国画大家吴冠中、黄永玉等人的重要作品，是康先生的收藏之宝。这些国画大师的作品就散落在公馆的某几间客房和餐厅里，使来宾有机会近距离地体会和感受中国文化的精粹与优

雅。健一公馆，让来宾真正体会到移步换景带来的每一次惊喜。

东方文化海纳百川，从来不乏对外来文化的兼容并蓄，健一公馆的内里配套设施便是中西合璧的大胆组合，红酒吧、雪茄吧的设置，是对现代生活方式的包容。中式设计的尺度恰到好处，雕花木床与西式的沙发和谐共存。"中式为体，西式为用"的精神在这里得到了良好的体现。

健一公馆作为闹市区中难得一觅的绿色与文化净土，拥有宁静雅致的住房环境。中式经典的客房设计，24小时管家服务等，这些都是健一公馆追求国际级中式精品酒店品质的具体呈现。如果不是亲临，也许很难想象徜徉在每间客房的私家院落中的悠闲惬意，或与三两知己于阁楼书房促膝而坐、指点江山的畅快淋漓，或独自于水系平台之上淡然地享受早餐的自在平和。

公馆还设有高级中餐厅为来宾呈现中华美食，涵盖中国四大菜系。16间中式顶级风格的包房更是客人独享的尊贵场所。在享受美食的同时饱尝满眼的大自然美景。简洁时尚的酒吧区呈现出中式元素与西式风格的完美融合，并拥有来自世界各地的丰富藏酒，无论阳光慵懒的午后，抑或静谧魅惑的夜晚，都可以来一杯鸡尾酒，浸润一段惬意时光。

健一公馆传承中国文化，着力打造世界级中式精品酒店，用心为所有客人提供一个在北京的家。

Just as an old Chinese saying goes, the style is the man, which means that we can know a man by judging from his style. If you want to get an understanding of Jianyi Mansion, you have to know Kang Jianyi, its owner, an admirer of traditional Chinese culture and a collector. From his point of view, the greatest pleasure and enjoyment in life come out of the perception of history from details of a historical item. Thanks to his working experience in auction, he holds distinctive views on art collection that Chinese oil paintings are like herbs, undertaking suspicion of a great speculation current, about which he is not very enthusiastic. On the contrary, Chinese ink painting is relatively more appealing, because the whole process lays emphasis on gradual infiltration into the paper from the beginning, which is bound to generate a wealth of extremely subtle changes of ink, that are used to emphasize the charm and to realize the pursuit of an artistic realm where the images are left without the physical appearance while the essence leads to the spiritual level. The experience in borrowing ideas from Qi Baishi, a master of traditional Chinese painting, furthered his preference for works of the late Qing Dynasty. Of course, Mr. Kang has been collecting a lot of works by Qi Baishi and creating works of art himself.

And now we can see why Mr. Kang desired for a boutique high-end Chinese mansion. It appears to him that the key to making a Chinese-style boutique hotel is to show the connotation of China's extensive and profound culture in details, rather than to simply accumulate a pile of Chinese elements.

With the context above, the essence and the wit of this space are naturally bound to come out.

Located in the west section of Red Scarf Park, the Jianyi Mansion boasts its open view to the lawn of 5,000 m^2 outside, which makes people relaxed and is really unparalleled luxury in Beijing where the value of land is extremely high. Therefore many top brands in the world like Mercedes-Benz, BMW, and Piaget have chosen to cooperate with Jianyi Mansion to hold large-scale press conferences.

Jianyi Mansion is mainly designed in Chinese style, with an elegant and quiet appearance, which reveals steadiness, a typical feature of architectures of the Republic of China, while the interior space is different. Inspired by the Forbidden City, Mr. Kang employs the traditional elements in the design of the mansion. The deep Chinese cultural background including issues such as solemnity, elegance, symmetry and order is thoroughly reflected in public spaces from the entrance lobby, the hall, and even the corridors to the guest rooms. Each corner is a piece of artworks, which oozes an inner sense of elegance and implicature, and each detail is the best expression of Chinese culture. In the lobby, wooden beams allow for an image of a steady, calm patio. On both sides of the wooden corridor the stone and the brick cavings strike you with a great surprise for Chinese handcrafts. It is stunning of the complexity and skills reflected by the wooden carvings in public spaces and the decorative wood carvings in guest rooms. However, all of these are not enough at all. Calligraphies, paintings and cultural works around the mansion are eye-catching, especially those masterpieces by Wu Guanzhong and Huang Yongyu, both of whom are masters of traditional Chinese painting and their works in some guest rooms and the dinning hall allow for a close contact to the essence and elegance of Chinese culture. Jianyi Mansion allows guests to experience the surprise at the shifting views as they go around.

The sea admits hundreds of rivers for its great tolerance, which is the charm of the oriental culture, tolerant and diverse. The facilities are a bold combination of the eastern and the western styles, like the red wine bar and the cigar bar, which an inclusion of modern lifestyle. The Chinese carved wooden bed keeps in harmony with the western sofa. The spirit has been well embodied that the Chinese style serves as contents while the western style functions as forms.

The mansion is a rarity of a green and quiet cultural environment in downtown, with the elegant and exquisite living environment. There are guest rooms designed in Chinese classical style and 24-hour services, all of which bring out the quality of a world-class Chinese-style boutique hotel. If you are not in the mansion, it is hard for you to imagine the leisure of the private courtyard in each guest room, the great happiness from friends-chatting or commenting on events domestic or international and the peace and tranquility coming about with the sitting on the waterfront platform.

In the high-end Chinese restaurant China's four major styles of cooking can be available, where 16 top Chinese-style boxes are for distinguished guests to exclusively enjoy the food and the beauty of nature simultaneously. The simple and stylish bar is a perfect combination of Chinese and Western elements with wines from all over the world. You can enjoy a glass of cocktail and relaxing time in a cozy afternoon or a quiet and charming night.

Jianyi Mansion is a project to inherit Chinese culture and tradition, to build a world-class Chinese boutique hotel, and to make a home in Beijing for those far away from their homes.

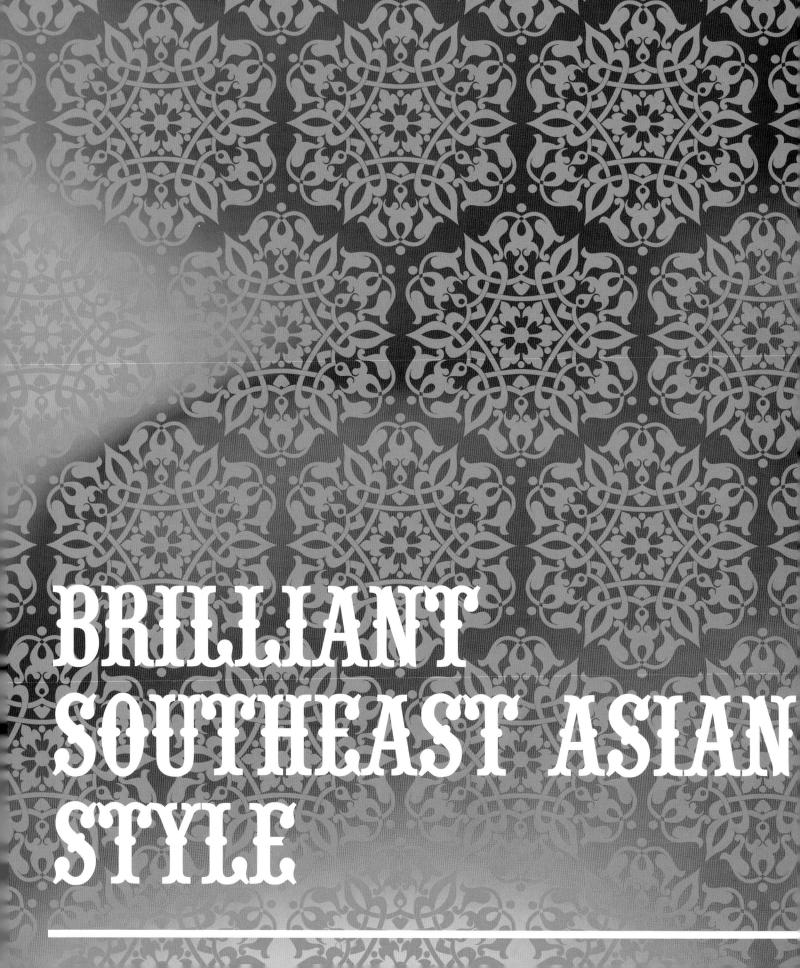

浓情东南亚

时尚撞上古典，迸发风格火花
WHEN FASHION MEETS CLASSICISM

- 项目名称：中国台湾中悦新天鹅堡
- 设计公司：岩舍国际设计事务所
- 设计师：林济民、陈韵如、吴家桦
- 面积：660 m^2
- 材料：大理石、柚木、木地板、进口壁纸、茶镜喷砂、锻铁栏杆、柚木实木扶手
- Project Name: Schloss Newschwanstein in Taiwan, China
- Design Company: Anta Design
- Designer: Lin Jimin, Chen Yunru, Wu Jiahua
- Area: 660 m^2
- Material: marble, teak, wood flooring, imported wallpaper, sand blasting tea glass, wrought iron railings, solid teak handrail

当时尚撞上古典，当优雅遇上阳刚，会产生一种怎样的效果呢？中悦新天鹅堡整个空间正好向我们诠释了这样的一种生活态度：不媚俗、不一味迎合潮流，充满了浓郁的文化气息并坚持自己独特的品味。

传统也好，现代也好，简约也好，繁复也好，高品位的各类元素可以相互融合出属于自己的风格，中悦新天鹅堡用事实证明，自己的风格才是最好的风格。

在一幅由抽象符号组成的现代艺术图像引领下，人们进入整体以低调奢华着眼的居家空间：四张两两相称的方形沙发围绕客厅，抱枕颜色各异，挑高的天花悬挂正方形吊灯，明亮的光线一泻而下，结合透亮的茶几及花束，组成宽敞大方的客厅环境。加上柚木海岛形地板，其极具质感的表面，让整个客厅透出传统的韵味。

餐厅以深色为主色调，有深灰的餐桌椅、深灰的落地窗帘以及深色的外墙景观。巨幅自然生物的油画作为主墙面装饰，为餐厅空间注入了原生态的色彩，并和客厅的主色调相互辉映，形成时尚简练的立面。

柚木实木楼梯连接了客厅、餐厅、走道三处主要空间，以铁灰色水波纹进口壁纸，搭配玻璃镜面展示柜及艺术品展示楼梯间。在灯笼式吊灯反射光源下，凝聚出优雅利落、品味卓然的视觉空间，在低调奢华中内含雅致内敛的氛围。而拾级而上的第一个着眼点就是休息室，一样挑高的天花，不一样的视觉颜色，几片芭蕉扇片镶嵌于主

负一层平面图 Basement Floor Plan

一层平面图 First Floor Plan

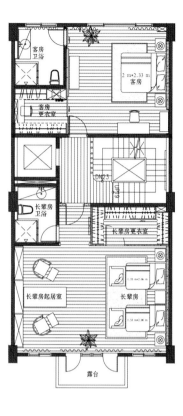
二层平面图 Second Floor Plan

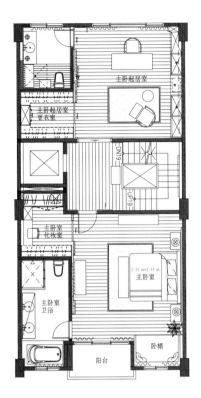

三层平面图 Third Floor Plan

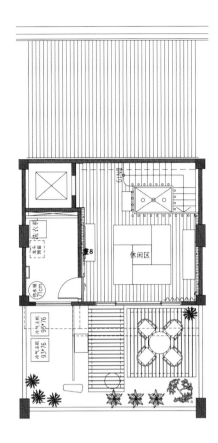

四层平面图 Fourth Floor Plan

墙，加上低矮的茶几坐垫，将休闲感表现得淋漓尽致。

拐个弯就是卧室了。设计师以不同的颜色配合作为设计主题，如阳刚气十足的朱古力色、感性的浅黄色、成熟稳重的银灰色。卧室空间的线条简洁而舒适，和床平行的右面多出一张极具现代时尚感的沙发，自然光线透过稀疏的木构天花洒落周围，这种细腻的设计于细节之处，尽显精致完美。加上巨大的乳白色时尚花纹地毯，诠释了房间的低调奢华，且让简约的房间透出优雅。

书房除具有阅览、休闲、娱乐的功能外，还在柜内设置了隐藏伸缩床架，可作为客房使用。清一色的木质构件、沙发、优雅的古典摆件、香醇的咖啡、乳白地毯结合特别制作的柚木地板，尽情表现个人的精神世界与修养。

When fashion bumps into classicism, and when masculinity meets elegance, what effect can be generated? The design of this project is the answer: no kitsch, no pandering to fashion, only adherence to its own unique taste that is full of cultural atmosphere.

Various high-grade elements ranging from tradition, modernity, simplicity to complexity make a unique style by mutual integration. This project is a good proof that only an individual's style can be the best style.

A modern art image composed of abstract symbols guides us into the low-key luxurious living space: a sofa set of four square sofas encloses the space; cushions are in different colors; the ceiling is raised high; the square chandelier hangs from the ceiling; and the tea table with a bunch of flowers is bright. A large and generous living room environment is thus formed. And the island-shaped teak flooring with its much textured surface displays a traditional flavor.

The dining room is mainly in a dark hue, with the deep gray tables, chairs, the curtain of the French window and the external wall. The giant painting of natural creatures serves as the decoration for the main wall. The pristine color is therefore injected into the dining room, and sets off the tone of the whole space, which contributes to a fashionable and concise facade.

The solid teak stairs make a smooth connection of the three spaces, namely the living room, the dining room and the passage. The staircase is displayed with its imported iron-gray corrugated wallpaper, glass mirror display cabinets and works of art all together. The reflection of the lantern Chandelier makes an elegant, neat, and eminent visual space. The stairs wind up through the low-key, luxurious, and somewhat elegant but still restrained atmosphere. Then the lounge would firstly strike your eyes with the same high ceiling and different colors. Several embedded banana leaves and the cushions on the low tea table most vividly describe a sense of leisure.

Around the corner is the bedroom. Different colors are carefully selected as the theme of the design, like chocolate, light yellow, and silver. The lines of the bedroom are concise and comfortable. The sofa parallel to the bed is highly modern and stylish. Natural light shines through the thin wooden ceiling, which is a piece of delicate design. The huge milky white carpet interprets the low-key luxury, thus the simple internal space becomes elegant.

The study is designed practically for reading, relaxing, and entertainment. A retractable bed hidden in the cabinet could change the study into a guest room if necessary. With the wooden furniture, the sofa, the elegant classical furnishing articles, the fragrant coffee, and the teak flooring which is specially covered with a milky white carpet, the atmosphere here is harmonious and conveys the owner's pursuit of self-cultivation and spiritual world.

大气格局，尽抒写意情怀
MAGNIFICENT MANSION FOR SPIRIT CULTIVATION

- 项目名称：泰国张公馆
- 设计公司：岩舍国际设计事务所
- 设计师：林济民
- 面积：550 m^2
- 材料：大理石、柚木、胡桃木、ICI乳胶漆、进口壁纸、茶镜喷砂、锻铁栏杆
- Project Name: Thai Zhang Mansion
- Design Company: Anta Design
- Designer: Lin Jimin
- Area: 550 m^2
- Material: marble, teak, walnut, ICI emulsion paint, imported wallpaper, sand blast tea glass, wrought iron railings

东南亚风格的材料天然而色彩艳美，看似矛盾却又散发着神秘而诱人的气质。虽然我们无法用三言两语将其完全描述清晰，但总有一些标志性的搭配让我们心领神会，例如镶以软垫的木质坐椅、泰式抱枕、精致的木栅格门、造型逼真的佛手、妩媚的纱幔等。当这一切和谐地兼容于一室时，我们便能准确无误地感受到那种东南亚的清雅、休闲的气氛。

本案是一座双连体的两层别墅，尺度开阔，各个功能区围绕着中心转角旋梯布置，使每一个功能区都拥有良好的观景面与通风采光格局。

公馆的玄关入口，虽然是以天然材料做框架，却分明有丝丝贵气自然流露。编织纹样的柚木大门，厚实而大气，两侧百屉拉柜上两尊青铜古董雕像守卫着家宅的平静安宁。

恢弘的会客厅采用对称式格局，没有花哨的装饰造型，通过利落的线条和精确的比例营造出庄重典雅但轻松怡人的氛围。客厅四周皆以柚木材料做框架，饰以方框与矩框的拼花图案，同样的拼花在木质门扇和茶色玻璃上反复使用，统一着空间的调性。紫色绸面的沙发，围绕着朱红方形大理石面的茶几，与不同款式的实木布垫坐椅共处一室，和谐中透露着细微的变化。围绕着沙发区，各式艺术品、古董摆件、民族风情画点缀着空间的各个角落，显露出"大户人家"卓尔不凡的品位与鉴赏力。

餐厅与厨房在公馆首层的另一边。长形餐厅拥有开阔的景观面，两侧可开门到达花园，中间的"八"字形落地窗前，摆放着餐边柜，人们可以端坐在餐厅中，就着美景入肴，尽享其乐融融的大家庭氛围。

卧室的设计沿袭了客厅的部分风格，深黄色的柚木材质围成的墙面、柚木地板和落地窗户，衬托着卧室的慵懒闲适。落地窗户两旁的壁橱里精心摆放着古陶古罐，灰白相间的靠垫堆叠出几许安然。设计师利用不同材质将单一的色彩解构为层次鲜明的局部，避免了过于单调的色块堆积。卧室一角造型简单的沙发椅配合造型简约的台灯和案桌，布置出了一处可供阅读和观景的休闲角落。

而书房摆脱了卧室空间的单调，代之以深木色，为的是打造一个凝神静气的平和空间。别出心裁的书柜设计、可同时供几个人上网办公娱乐的电脑桌及简洁的空间融合了现代设计的理念。白色的沙发上随意摆放的杂志，给人带来了休闲轻松的感觉。

如斯美宅，总有一个角落，让人"偷得浮生半日闲"。

The Southeast Asian style is composed of natural materials and bright colors. Seemingly contradictory, it is mysterious and seductive. Although we can't describe it thoroughly in a few words, we can understand it through a number of significant elements, such as wooden seats with soft padding, the Thai–style pillows, the exquisite lattice wooden doors, the Buddha's hands and soft gauze, all of which fit into the interior space and create an elegant and casual Southeast Asian atmosphere that can be perceived.

The building is a two–story two family house. Open and large, all functional areas are arranged around a central revolving ladder, so that each area opens to a fine view. In addition, it creates good ventilation and lighting.

The entrance portal is clearly noble and elegant although it is a framework made of natural materials. The thick, solid and splendid teak gate is weaving–patterned, on whose sides the cabinets stand with two antique bronze statues, guarding the peace and the tranquility of the house.

The parlor is broad and symmetrical without any fancy decorations. It is designed with concise lines and precise proportion to create a dignified, elegant, relaxing and pleasant atmosphere. Around are the teak frames with square and rectangular patterns, which are also used on wooden doors and tinted glass to be consistent. The purple silk sofas around the red square marble tea table are accompanied by the solid wooden pads of different types, which are harmonious but subtly variable. Around the sofa area, all kinds of artworks, antique ornaments and folk paintings are displayed, revealing the extraordinary taste and appreciation of a rich family.

The dining room and the kitchen are located on the other side of the first floor. The long dining room enjoys a broad view, with both sides accessible to the garden. In the middle is a French window, in front of which is a kitchen cabinet. People could enjoy their meals and an exhilarating family atmosphere while appreciating the beautiful views outside.

The style of the bedroom is partly a continuation of the living room. The wall decorated with deep yellow teak, the teak floor and the French windows merge into the leisurely and relaxing ambience of the bedroom. On both sides of the window are the closets, where ancient potteries and pots are well–stocked, and cushions chequered with gray and white make a steady sense. With different materials, the designer enriches the monochrome with clear gradations of color, avoiding the monotony of color. With a simple sofa and a lamp, as well as a table, the corner makes a leisurely place for reading and viewing.

However, the study is designed in deep wood color, different from the monotonous style of the bedroom, to create a tranquil space for meditation. The bookcase is designed uniquely. The computer desk is large enough for several people to surf the internet at the same time. The simple space is designed in a modern concept. The magazines on the sofa reveal a feeling of relaxation and idleness.

There is certainly a corner for you in this mansion where you can snack a little leisure from a busy life.

自然东方，生活长青
THE NATURAL EAST, THE EVERGREEN LIFE

- 项目名称：长青湾自然东方样板间
- 设计公司：北京睦晨风合艺术设计中心
- 设计师：陈贻、张睦晨
- Project Name: Show Flat of the Natural East of the Evergreen Bay
- Design Company: Beijing Muchenfenghe Art & Design Center
- Designer: Chen Yi, Zhang Muchen

此套样板间风格定位为自然东方，无论设计理念，还是材质、色调、大量植物的合理运用，都完全切合了人们向往低碳自然的生活态度，也成全了北方人对南方生活环境的向往和遐想，用设计语言表达了人们的心思，符合现代人的需求。

设计师在该样板空间中运用了纯粹的、地道的并且是现代的体现东南亚度假胜地风格特色的设计语言和元素，如木质格栅、天花木梁、花格吊顶，以及泰国和印尼的皮制和藤制家具与配饰。在空间营造上，样板房继承了自然、健康和休闲的特质，大到空间打造，小到细节装饰，都体现了设计师对自然的尊重与协调，以期营造出东方的诗意空间和浓郁舒适的浪漫情调。住在这里能让人心情充分放松，回到舒适和自然平和的心情状态上来，感受文化沉淀的厚度和深度。

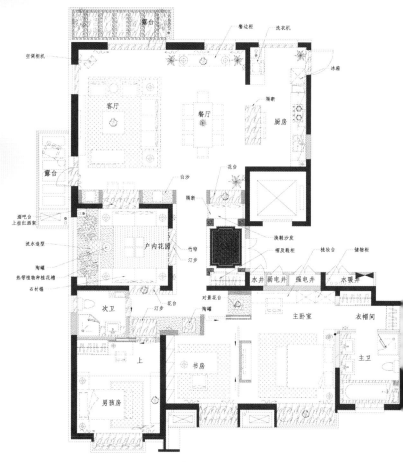

平面布置图 Layout Plan

对于空间的改造，设计师陈贻没少动脑筋，可谓"大动干戈"，将原本枯燥的格局打造成极具特点的室内空间。5 m²的室内露台被打造成户内花园，石材质的沙发、绿色大植物，有种东南亚蜜月的气氛，人们在这个空间里似乎身处遥远的大自然当中，满足了人们对自然的向往。设计师对室内露台与客厅之间的窗户进行了合理的运用，设计了一个吧台，上方有酒架，两边增置了凹龛，特别挑选的木制摆件，为空间增色不少，两个空间顿时增添了一丝神秘感，并且遥相呼应。与此同时，客厅门的另一侧对应地也添置了一处凹龛，以达到视觉上的平衡效果，也为玄关处增大了空间，格栅后与其对应的是石材质的沙发，多置一处景观，多一个功能，又多了几分味道。厨房采用开放式处理方法，将原有隔断开，也打开了视野，体现主人追求舒适的生活态度。书房与主卧室进行了大的改动，这两个空间采用推拉门的形式互相沟通，并采用了日式的元素。书房是进入主卧的必经之路，原有的门被封住，改造成了一处凹龛摆台，用东南亚风格的饰品来点缀，书房与儿童房让出的空间配一陶罐，置于白色鹅卵石上。此情此景富有私密的异域情调，从客餐厅方向望过来，又是一处怡人的景色。

This show flat features a natural and oriental style. The design concept, the employment of materials, colors, and plants completely meet people's yearn for a natural and low-carbon style, and satisfy the northerners' desire for a southern living environment. The design language is a good expression of people's minds and meets their needs.

The pure, authentic design language and elements which present the style of modern Southeast Asian resorts are employed in the show flat, such as the wooden grilles, ceiling beams, lattice ceiling, the leather and rattan furniture and accessories of modern Thailand and Indonesia. The space reflects the characteristics of nature, good health and leisure. The entire design and the detailed decoration show a respect for nature and harmonization, creating a poetic space and rich oriental romance and comfort. After full relaxation here people can return to peace and tranquility with an experience of the profound cultural background.

Chen Yi, the designer, particularly put emphasis on the transformation of the space, converting the original and dull layout into one with unique features. The indoor terrace, which has an area of 5 m^2, is changed into an indoor garden with stone sofa and green plants, which provides the occupants with a kind of Southeast Asian honeymoon atmosphere, and allows them to feel as if in the far-away nature. It reflects people's desire for nature. The designer made good use of the window between the indoor terrace and the living room. Around the window there is a bar, above the bar there is a wine rack, and on both sides of the wine rack are the niches with the specially-selected wooden ornaments, which brighten up the space. The spaces of the indoor garden and the bar thus become more mysterious and correspond with each other. Meanwhile, the other side of the door of the living room is also equipped with a niche, in order to achieve a visual balance and expand the space of the doorway. Behind the grid is the corresponding stone sofa, which brings another scene and another function, adding another taste to the space. The original partition of the open kitchen is removed for an extended view, meeting the owners' pursuit of a comfortable life. The study and the master bedroom have been greatly transformed. The medium between is a sliding door of Japanese style. The original door of the study, which was closed, was converted into a niche decorated in Southeast Asian style. The left space of the study and the children's room is decorated with a jar on the white pebbles, which offers a private, exotic and pleasant view from the dining room.

光影游弋，自在闲庭
THE DYNAMIC LIGHT, THE LEISURELY COURTYARD

- 项目名称：单陛二层样板间
- 设计公司：北京睦晨风合艺术设计中心
- 设计师：陈贻、张睦晨
- Project Name: 2F Show Flat
- Design Company: Beijing Muchenfenghe Art & Design Center
- Designer: Chen Yi, Zhang Muchen

本案整体设计偏于现代风格，而适度的东方元素低调地穿插其中，并不喧宾夺主，带有东南亚风情的横木栅栏条贯穿了整间样板房。纵观全案，设计师擅长挑选精致的材质，用层层光影营造出空间的美感和价值，令人身处其中心旷神怡，惊喜无穷。设计师还擅用镜面材质，充分展现了对材质及空间运用的自由度。全室多处以匠心独具的设计语汇，重新诠释材质的各种可能性。

客厅以简洁不失典雅的设计风格，表现大气风范。天花和沙发背墙上的镜面材料，丰富和扩大了空间的光影层次；大面积的电视墙面以石材铺陈，两边则延续了横木条栏的装饰要素。宽大的落地窗，可以"贪婪"地吸收清晨的第一缕阳光，而经由室内镜面材质、石材等的一系列反光折射，自然光与人工光混合交杂的斑驳光影，星星点点地游移在室内的各个角落。

本案的设计也侧重于空间感的确立，设计师探究空间里的转折与区域动线的规划，借由虚实交错的景况贯穿，增加趣味性及丰富性。房间整体以开放的格局尺度、对比的彩度作规划基调，使空间形成纯粹、透明、舒缓的宁静氛围，增加了区域之间的互动，光影的变迁让室内宁静、丰富而多样化。

The show flat is designed in a kind of modern style, and oriental elements are moderately integrated into it. The crossbar fence in the Southeast Asian style is applied throughout the entire show flat. The designers are experienced in selecting exquisite materials and displaying the beauty and value of the space with light and shadows on different gradations, which brings you with refreshment and surprises. The wise employment of mirrors reflects the diversity of ways in which the designers make use of the material and their sound understanding of the material and space. The entire show flat is designed uniquely in every part, showing to what extent the materials can be used.

The style of the magnificent living room is simple yet elegant. The mirrors on the ceiling and on the background wall of the sofa multiply the gradations of light and shades. The large TV wall is made of stone, both sides of which continues to use the decorative elements of the crossbar fence. The first ray of morning sunlight can get through the broad French windows. With a sequence of reflection and refraction by the mirrors and stones inside, a mottled mixture of natural light and artificial light scatters everywhere.

Additionally, the design of the show flat focuses on the establishment of a sense of space. The designers study the transitions between areas and circulation planning, adding fun and richness to the space by a mixture of the real and virtual scenes throughout the show flat. On the whole, it is planned on the basis of an open layout and a contrast between tones, creating a pure, transparent, relaxing and quiet atmosphere and increasing interactions between areas. The changes of light and shadows enrich and diversify the tranquil space.

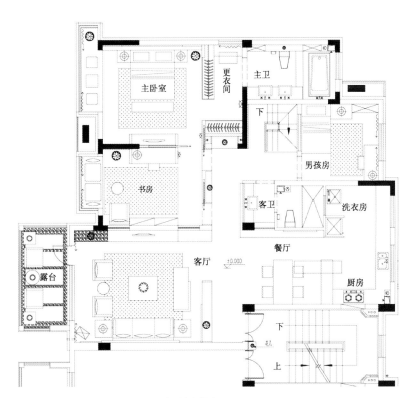

平面布置图 Layout Plan

深紫幽蓝，浓情美居
THE BEAUTIFUL RESIDENCE OF PURPLE AND BLUE

◆ 项目名称：万科样板房
◆ 设计公司：北京睦晨风合艺术设计中心
◆ 设计师：陈贻、张睦晨
Project Name: Vanke Sample Flat
Design Company: Beijing Muchenfenghe Art & Design Center
Designer: Chen Yi, Zhang Muchen

空间整体设计以木材为原材料，暗红与浅黑交错的格子地板、深棕色雕花木椅、暗红色木质家具、交叉菱形长窗、长方形排列而成的天花，无一不恰到好处地彰显了木材特有的稳重、古朴的气息和丝毫不显笨重的东南亚民族风格。

房间细节处民族气息浓重，如螺旋盖帽状的吊灯、墙壁上的围巾式挂饰、暗花窗帘、镂空手编桶状凳子，就连地毯也颇具东南亚风味。在以暗色调为主的空间里，最显眼的莫过于那白色的沙发，它静静地躺在那儿，丝毫不逊色于边侧与周围环境相融合的一张木质躺椅。奢华优雅的视觉感，不突兀也没有喧宾夺主，它的存在让古典风十足的装饰多了一些现代风格，使整体装潢显得更现代，却又不失民族特色。

各个空间相通的一大好处是可以让人一走进玄关就对整体一目了然。把各个大小空间连结在一起的枢纽是延伸客厅而下的廊道，那里别有一番风趣：棕色的阶梯式天花、棕色的镂空木质格子窗、暗黑的地板，衬以窗外透进来的缕缕光线，最夺目的是从楼梯口的盆栽里伸展而出的那几枝翠绿，和整个廊道空间的色彩相映成趣。

最能体现东南亚风格的非卧室莫属了。主卧室选用了深色的木质天花，金色木质墙面结合了光线的变化，深灰色的床垫和抱枕，深灰色的落地窗帘，随意拖落到地上的具有泰式风格的床帘，其漫不经心的造型，多了一种被风一吹就能随之起舞的姿态，使卧室里飘逸着一份轻盈，创造出一种雍容华丽之感。

餐桌上方的铜镜作为装饰似乎也是一绝佳物品，在灯光下显金色的铜制边框，不在灯光范围内则泛着银色的光芒。尤其是铜镜里映照出来的那一抹绿，在步入餐厅的瞬间便很容易地被人注目。

The Flat is mainly made of wood. The grid floor interlaced with dark red and light black, the brown carved wooden chairs, the dark red wooden furniture, the overlapping diamond windows, and the ceiling composed of an array of rectangulars appropriately highlight the steady and primitive feature exclusive to wood and set off the Southeast Asian style that is by no means dull.

Its details are rich in the folk breath, with the screw–shaped chandelier, the scarf–style ornaments on the wall, the floral patterned curtains, the hollowed–out hand–woven barrel–shaped stools, and the Southeast-Asian-style blankets. The dominant feature of the dark space is the white sofa, which is static and as attractive as the wooden lounge on the side which fits in with the surroundings. Visually, it's luxurious, elegant and moderate. It makes the ornaments of ancient style and the entire decorations more modern while maintaining the ethnic features.

Thanks to openness of the spaces to each other, the entrance provides a panoramic view. The interesting corridor extending from the living room serves as a hub which connects the spaces of different sizes. There is a brown ladder–shaped ceiling, hollowed–out brown wooden lattice windows and rays of light spilling on the dark floor through the window. The most eye–catching scene is the green leaves of the potted plants at the stairs, which corresponds with the color of the corridor.

It is the bedroom that most embodies the Southeast Asian style. The master bedroom is equipped with the dark wood ceiling, the golden wooden wall which changes with light, the dark gray mattress and the pillows, the dark gray curtains, and the Thai–style bed curtain which casually swags on the ground as if it is ready to dance once there comes a breath. The sense of lightness in the air comes along with its magnificence and elegance.

Above the dining table is a wonderful item, a decorative bronze mirror. The bronze rim glitters with gold in the light while the other parts of the mirror out of the light glow with silver. The twinkle of green from the mirror is the most special and the attractive view within the whole dining space.

回归自然的东南亚风情
RETURN TO THE NATURAL SOUTHEAST ASIAN CHARM

项目名称：东莞万科·棠樾澜山居样板房
设计公司：深圳市昊泽空间设计有限公司
设计师：韩松
面积：88 m²
材料：戈壁沙漠石材、橡木面板
Project Name: Sample Flat of Vanke Tangyue Lanshan, Dongguan
Design Company: Shenzhen Haoze Space Design Co., Ltd.
Designer: Han Song
Area: 88 m²
Material: stone, oak panel

走进房间，一种雅致温馨的感觉扑面而来，虽是东南亚风格的设计装潢，但摒弃了繁复华丽的装饰，整个空间给予人一种回归自然的想象。

木石结构、砂岩装饰、墙纸、浮雕、木梁、漏窗……这些都是东南亚传统风格装饰中不可缺少的元素。多功能的木柜在客厅占了一面墙的位置，不可思议的是，木柜里不但有饰品、茶具、盆栽，还有供娱乐用的DVD音响。一袭浅灰色雕花地毯延伸开来，与桌子上的绿色盆栽、芭蕉叶状装饰构成的吊灯以及天花的橘黄形成了强烈的反差，展现出一种对比美。

为了方便出入并使空间更宽敞，设计师将客厅、餐厅和卧室巧妙地组合，将三者依次铺张开来，使其同在一条水平线上，一同构成了相互渗透的活动区域，把整体空间延伸得足够宽敞。这里无论是地面、天花还是墙壁，都处理得更加接近自然，以优雅大气为主。木质餐桌、竹制椅子、木质暖色调的卧室层叠门和墙面雕花纸纹造型延续统一的风格。偌大的主卧露台非常方便，提供了一个和自然接触的空间。整个空间以简约线条分割，代替一切繁复的设计，不需矫揉造作的材料却营造出度假感。

卧室也将东南亚风情演绎得旖旎浪漫。一袭轻柔的纱幔环绕出梦幻般的睡眠区，几个五颜六色的泰丝垫枕漫不经心地点缀在床具上，给人度假般轻松自然的感觉。

当你沉浸在东南亚风格古朴自然的空间里时，就如同沐浴着阳光雨露般的清新自然。

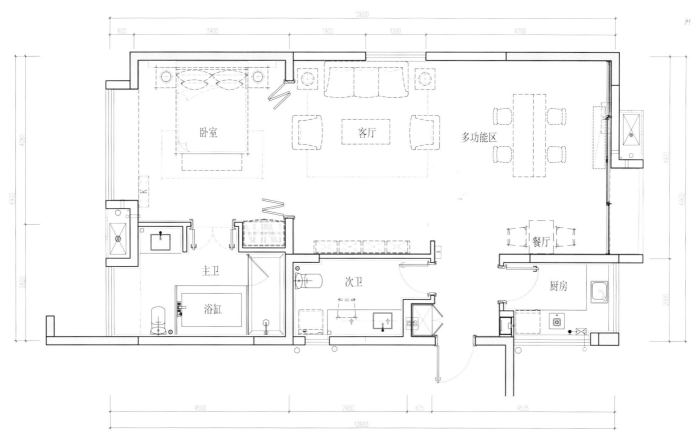

平面布置图 Layout Plan

What first strikes you out of the space is an elegant and warm feeling. Despite the application of design and decoration of Southeast Asian style, this project allows for an imagination of returning to nature with the rejection of complex or ornate decoration.

Essential elements of traditional Southeast Asian style of decoration are wood and stone structure, sandstone decoration, wallpaper, relief, wooden beams, and grilles. The multi-functional wooden cabinet in the living room occupies a wall. Incredibly there are accessories, a tea pot, potted plants, and DVD appliances for entertainment in the cabinet. On the ground is laid a light gray carved carpet. The green potting and the droplight composed of banana-leaf-like modelings make a strong contrast with the yellow ceiling, showing a contrasting aesthetic.

The subtle combination of the living room, the dining room, and the bedroom makes the space convenient and broad when the three different spaces spread one after another in a horizontal line and form activity areas that are open to each other, extending the whole space to be spacial enough. The ground, the ceilings and the walls are relatively natural and dominantly elegant and generous. The wooden dining tables, the bamboo chairs, the stacked wooden bedroom doors in warm hue and the carved wall fabric are in a continuous tone. The huge master bedroom terrace is an access to nature. The whole space is partitioned by simple lines instead of complicated design, and successfully creates a sense of resort without any artificial materials.

Additionally, the bedroom is a place to interpret the romantic charm of Southeast Asia. A gauze curtain encloses a dreamy sleeping area, and several colourful Thai silk pillows are scattered on the bedding, bringing out a relaxing resort-like feeling.

Indulging yourself in this ancient, pristine and natural space of Southeast Asian style, you would feel fresh and natural as if bathed in the sun or in the rain.

花果飘香，浓情雅居
THE ELEGANT RESIDENCE IN MOOD AND FRAGRANCE

◆ 项目名称：武汉·保利中央公馆
◆ 设计公司：广州市韦格斯杨设计有限公司
◆ 设计师：区伟勤、陈正茂
◆ 撰文：陈正茂
◆ 材料：黑白根、爵士米黄、文化石、木纹石、木饰面、木地板、席面壁纸

◆ Project Name: The Poly Central Residence in Wuhan
◆ Design Company: GrandGhostCanyon Designers Associates Ltd.
◆ Designers: Ou Weiqin, Chen Zhengmao
◆ Text: Chen Zhengmao
◆ Material: nero Margiua marble, regal beige, culture stone, serpenggiante, wood finishes, wood flooring, mat-patterned wallpaper

平面布置图 Layout Plan

本案以年轻白领及社会中层精英为主要客户，整体空间以黑色、米黄色石材的装饰为主，使空间体现其强烈对比，更以木色作为空间主色调，视线所及之处都散发着浓郁的东南亚风情。

进入客厅，墙身大面积搭配使用黑白根石材与米黄雕花石片，稳重中透着精致，而墙身悬挂的黑底金边的装饰品如花朵一般绽放，共同构筑出一派异域风情。客厅中央垂挂的一簇水果形状的吊灯，不禁让人想起亚热带气候下的果实累累的丛林植物，成为空间的点睛之笔。室内大量采用木料与棉麻材质的家具，其丰富的质感将东南亚的淳朴与自然通过视觉和触觉传达到住户的心里。

餐厅里，由石片拼接的具有岩石般粗糙效果的装饰墙上，一面用木棍拼成的具有光芒四射效果的艺术镜面绽放出太阳般的热情。镜子前面，木质雕花艺术案几上有一尊端坐的佛像，面目沉静而虔诚，令人不禁肃然起敬。

卧室的设计中丰富的色彩与细节蕴藏着居住者对生活的无限热爱。主人房以咖啡色系配橙红的落叶墙纸，在沉稳宁静中显现对激情的向往；女儿房在床头墙身以木质雕花摆件呈现出精致美感。

整个项目空间不大，却以丰富的层次、饱满的色彩和亲切的质感，共同构筑出一个洋溢着东南亚风情的气质之居。

The space, targeting at young white collars and the middle-level elite, is dominated by the application of black and beige stones to reflect a strong contrast. The space is mainly in wood hue and oozes a strong Southeast Asian sense as far as your eyes could reach.

A large area of walls of the living room is decorated with nero Margina marble, accompanied by carved beige stones, to reveal a steady and exquisite sense, as well as the golden-rimmed decorations on the wall with backdrop like flowers in bloom, all of which form an exotic style. In the center of the living room hangs a fruit-shaped chandelier, reminiscent of the fruitful sub-tropical jungle and thereby serving as the highlight of the whole space. All furnishings of wood, cotton and linen materials in the interior space convey the Southeast Asian sense of simplicity, primitivitsm and nature into the occupants' hearts by their rich texture through visual and tactile senses.

In the dining room stands a decoration wall split-jointed by stones to make a rock-like robust effect, on which there is an artistic mirror with the aid of crabsticks to generate a bursting-out effect, revealing the passion like that from the sun. In front of the mirror is a long wood carved table, on which sits a Buddha status, which looks quiet and devout, and then deeply from your heart comes out sincere respect.

As for the bedroom, an abundance of colors and details indicate the love for life of the occupants. The master bedroom in brown hue matches orange leaf wallpaper, reflecting the desire for passion from the calm and tranquility. Carved ornaments on the wall behind the head of the bed show a sense of delicate beauty in the daughter's bedroom.

Despite the limited footprint of the space, the project forms an elegant estate of Southeast Asian style with rich layers, colors and accessible texture.

木色倾城，闲情逸居
WOODEN LEISURELY RESIDENCE

- 项目名称：武汉拉菲中央首席官邸B
- 设计公司：广州市韦格斯杨设计有限公司
- 设计师：区伟勤
- 面积：420 m²
- Project Name: Wuhan Rafi Chief Residence B
- Design Company: GrandGhostCanyon Designers Associates Ltd.
- Designer: Ou Weiqin
- Area: 420 m²

本案为五层联排别墅，以现代东南亚的装饰手法作为户型的切入点，突出休闲的生活情调，以求摆脱忙碌的都市生活。

客厅是全案的精华所在，双层挑高的空间气势开阔。临客厅的一面，上下两层的木质隔扇门如屏风一般按序排开，营造出质朴、大方和开阔的空间气场。客厅背景墙以黑洞石铺排，悬挂红色挂画，显得端庄优雅。客厅对面为景观餐厅，其独立的采光玻璃大门，将室外美景延入室内，为餐饮更添情趣。

天然材质作为东南亚装饰风格中不可缺少的一个元素，在本案中被运用得入木三分。木质材料的运用无处不在：木质天花板、木质地板、木质门和木质墙壁，加上垂挂的刻有浮雕花纹的吊灯，形塑出空间的温暖质感。室内还将黄洞石、黑洞石、板岩、席纹墙纸这几种粗糙材质进行巧妙组合，把它们运用到室内的各个角落，形成了有趣的对比与丰富的层次。

设计师还运用现代设计手法，对东南亚木格元素进一步提炼，营造悠闲、地道的东南亚度假感受。香薰、花瓣和鲜艳色彩的点缀，给人以嗅觉和视觉上的冲击。

本案运用材质将空间营造出接近大自然的意境，希望能将度假的休闲心情融入到生活中，给住户一个专属的度假空间。

The project of a five-story townhouse is decorated in a modern Southeast Asian style to highlight a mood of leisurely life and to be free of the burden of the hectic urban life.

The generous double-height living room is the focal point of the whole space. On the side by the living room, the wooden sliding partition doors on two floors spread in order like a screen and generate a simple, generous and open sense. The backdrop is made of black travertine, on which hangs an elegant red painting. Opposite the living room stands the landscape dining room with a separate glass door, which introduces the exterior landscape into the interior space, adding fun to dining.

As an indispensable part of the Southeast Asian décor, natural materials are employed to the utmost: wooden ceilings, wooden flooring, wooden doors, wooden walls, and chandeliers engraved with relief pattern combine to form a warm texture in the space. Rustic materials in each part of the interior space, like yellow and black travertine, slate, mat-patterned wallpaper combine strategically, forming an interesting contrast and creating rich gradation.

The designers have refined the element of Southeast Asian wooden lattice in modern ways of design, thereby accomplishing a concise, relaxing and authentic Southeast Asian resort experience. The aromatherapy, flower petals, and bright colors are exerting an impact on smell and vision.

The project creates a realm close to the nature by materials to integrate a relaxing mood in resort into life and to provide an exclusive resort for the occupants.

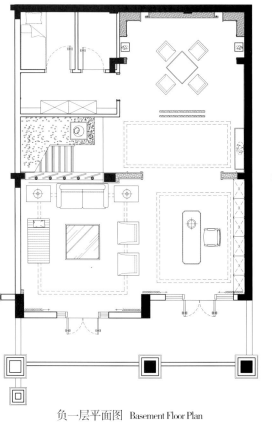
负一层平面图　Basement Floor Plan

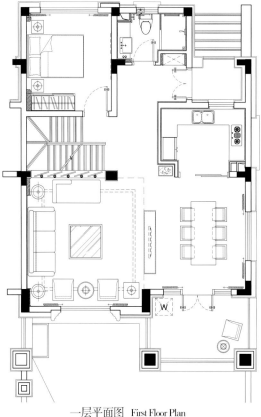
一层平面图　First Floor Plan

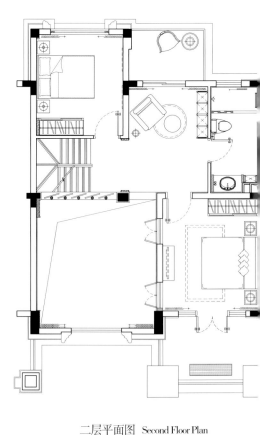
二层平面图　Second Floor Plan

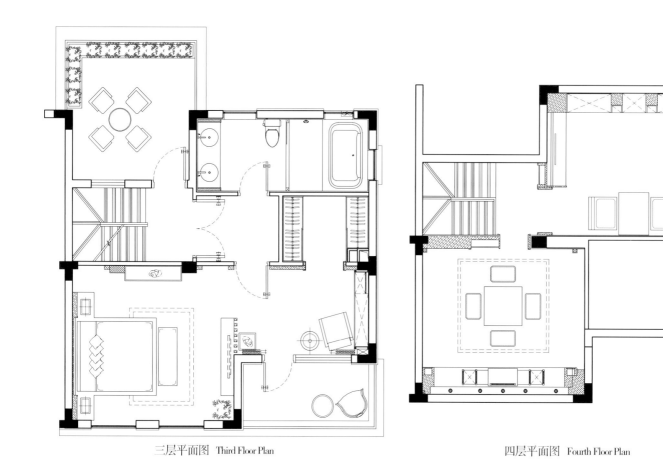

三层平面图 Third Floor Plan

四层平面图 Fourth Floor Plan

恋上东南亚，回归大自然
FALL IN LOVE WITH SOUTHEAST ASIA, RETURN TO NATURE

- 项目名称：长沙申奥美域
- 设计公司：湖南自在天装饰公司
- 设计师：杨啸
- 面积：140 m²
- 材料：原木、布艺、插花、花器
- Project Name: Shen Ao Land
- Design Company: Hunan Free-sky Decoration Design Engineering Co., Ltd.
- Designer: Yang Xiao
- Area: 140 m²
- Material: timber, fabric, flower arrangement, flower pot

在东南亚风情中，自然、原始的回归感是其主要特点之一。带有丝丝光泽的绿色窗帘与地毯将整个客厅打造得富有浓郁的热带气息，令人仿佛置身于热带雨林中，周围充盈着大自然的美妙音乐与阵阵青草香味。

由玄关通往客厅的通道，其导向性十分明确。整个客厅空间以深咖啡色作为基调，迎合了东南亚风情中原木家具的特点。同时，利用鲜艳的黄绿色与之搭配，融合了许多富有形状与视觉变化的图案，整个空间丝毫没有因深色而造成的沉重感，反而多了许多大自然的明亮舒畅与原始气息，令整个客厅洋溢着勃勃生机和天然风情。

与客厅相邻的是餐厅，造型特别的家具设置与灯光营造令整个餐厅显得精致小巧，富有浓厚的生活情趣。设计师还在背景墙上设置了银镜造型，这不仅点亮了空间，还使整体稳重却带有变化。虽然棕色会带有沉重的厚实感，但随着红色椅垫与晕黄灯光的渲染，整个色调显得清爽了许多。伴随着棕色壁纸、窗帘的诸多色泽变化，整个餐厅充满神秘气息，富有浓郁的浪漫情调。

卧室采用东南亚风格的四柱式床，白色的轻纱悬垂下来，布艺都经过精心设计，全手工打造，做工精致，所有的陈设配饰丰富而充满着质感，色泽大胆出位，但极具生活情调。

红色、绿色、蓝色、金色、白色等诸多色彩在同一空间当中演绎出丰富的变化，而配饰做主的空间更充盈着层次与厚度。

The Southeast Asian style mainly features nature and originality. With the shiny green curtains and carpets, the entire living room is filled with a tropical atmosphere, which allows people to feel as if in the tropical rain forest with the natural sound and the fragrance of grass around.

The passage from its entrance to its living room is clearly-oriented. The living room is mainly in dark brown, which matches the style of the Southeast Asian wooden furniture. At the same time, with the bright yellow-green color and the patterns in different shapes integrated into the design, the entire living room has become more naturally bright, comfortable and primitive and has been filled with an animate and natural ambience, without the dullness from the deep color.

Adjacent to the living room is the dining area, where the uniquely shaped furniture and lighting make the space appear compact and full of fun. A silver mirror modeling is set on the background wall and lights up the space, so that it is changeable to some extent while maintaining its stability. Although the brown color would bring out a sense of dullness, the entire tone seems much lighter with the red cushions and the yellow light. With various gradations of brown on wallpaper and curtains, the entire restaurant abounds with a mysterious and romantic atmosphere.

The bedroom is equipped with the typical Southeast-Asian-style bed with four posters. The white yarns drops and the hand-made fabric are exquisitely and carefully designed. All of the furnishings and accessories in the bedroom are of rich texture. Meanwhile, the colors of it are bold but show the fun of life.

Colors like red, green, blue, gold and white express a wealth of changes, while the space where accessories play a dominant role displays a sense of gradation and thickness.

别样混搭风，闲情东南亚
THE TREND OF UNIQUE MIX AND MATCH, THE LEISURE OF SOUTHEAST ASIA

◆ 项目名称：武汉住宅
◆ 设计公司：武汉梵石艺术设计有限公司
◆ 设计师：张瑾
◆ Project Name: Wuhan Residence
◆ Design Company: Wuhan Fanshi Yishu Design Limited Company
◆ Designer: Zhang Jin

本案以现代的手法诠释东南亚的装饰元素，把东南亚风格的原汁原味与现代风格的简单素雅自然衔接。

全屋以泰柚木色为中心，家具全是代表着奢华唯美的泰式家具，空间中最典型的特征是各类装饰与各类家具相互呼应。如客厅以电视背景墙上的巨幅石材拼花和柚木沙发相互辉映，沙发背景墙采用天然石材组合成抽象图案。又如同色系的刷漆壁布使餐桌色调在墙面上得到了延伸，又略带古禅韵味，同时这种韵味又在楼梯的上空得到了展现。

空间的局部装潢不逊色于整体，门厅以古朴自然的木色为主，洞石马赛克拼花地面，休闲舒适。圆拱形的门洞设计表现泰式风情，地中海元素的天花板设计，营造了在海边的浪漫气氛。加上泰柚木色和白色组合沙发，配上马赛克拼花的电视背景墙，用中式元素表现泰式风情，显得十分完美协调。

餐厅被设计于电视背景墙的反面，这种设计不仅在视觉上显得通透，而且给人很好的视野感。餐厅家具同样都是泰式风格的，很有情调，材质和花纹与厨房网格梭门呼应，很有整体感觉。

卧室的床，第一感觉就是舒适而宽大，木质框架表现出古典韵味。父亲房的床背景墙以金镜镶嵌泰柚格栅，油画古朴而气势不凡。母亲房的床背景墙采用艳丽的孔雀图案。此外，窗帘和软装饰都经过精心配搭，随意地放上一张造型独特的藤椅，造就了一间个性化的卧室。

除此之外，三楼会客厅完全是地中海风格的装潢。比较典型的地中海颜色的搭配即是蓝与白的搭配，空间门框、窗户、椅面都是蓝与白的配色。加上混着贝壳和细沙的墙面、小鹅卵石地面、拼贴马赛克、金银铁的金属器皿，蓝与白不同程度的对比与组合发挥到极致。

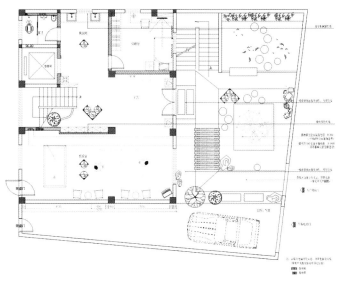

一层平面图　First Floor Plan

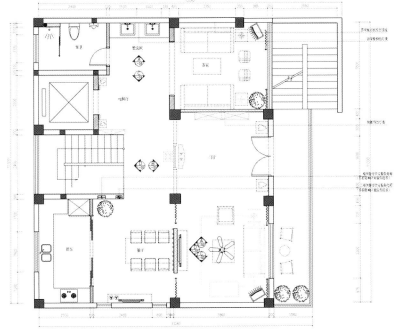

二层平面图 Second Floor Plan

With modern techniques, the project represents the decorative elements of Southeast Asia, smoothly combining the original Southeast Asian style with the simple and elegant modern style.

The whole house is mainly in Thai teak hue. All furniture is Thai-style and luxurious. The most typical characteristic of the room is that all kinds of its decorations could match the corresponding furniture. For example, the huge stone collage on the TV wall matches the teak sofa, the background wall of which has abstract pattern composed of natural stone; the painted wallpaper is of the same color as the dining table and slightly reveals the style of ancient zen, which could also be sensed in the air above the stairs.

The decorations of each part of the house are as splendid as the whole. Its foyer is mainly in classic and natural wooden hue, and the parquet there is travertine mosaic, which is casual and comfortable. The arched openings are of Thai style, while the Mediterranean-style ceiling creates a romantic atmosphere reminiscent of the beach. The hue of Thai teak and the white combination sofa, along with the mosaic TV wall perfectly present a harmonious Thai style with Chinese elements.

Behind the TV wall is the dining room. This design not only allows you to see through the whole space, but also provides you with a good view. The furniture there is also of Thai style. All the materials and patterns fit in with the sliding grid door of the kitchen, which appears as a whole.

The large bed gives you a sense of comfort, and its wooden frames are classical and charming. The gold-rimmed mirror is inlaid with Thai teak grille on the background wall of the bed in the father bedroom, and the oil painting there is simple but impressive, while the background wall of the bed in the mother bedroom has beautiful patterns of peacocks. Besides, the strictly-selected curtains, interior decorations as well as the unique cane chair make a personalized bedroom.

In addition, the sitting room on the third floor is entirely decorated in Mediterranean style. The typical Mediterranean color is a match of white and blue, thus the frames, windows and seats in the space are all in white and blue colors. Additionally, with the wall decorated with shells and sand, the floor of small pebbles, the collage mosaic and the household utensils made of gold, silver or iron, the combination and contrast between blue and white reach the upmost level.

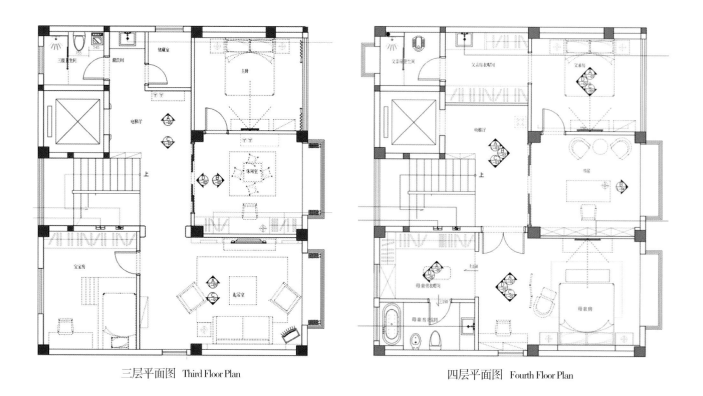

三层平面图 Third Floor Plan

四层平面图 Fourth Floor Plan

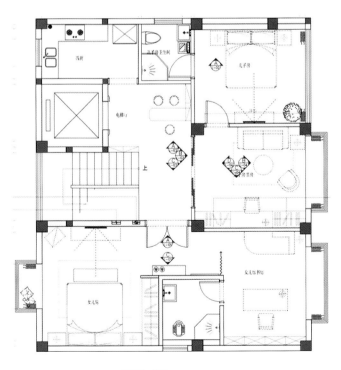

五层平面图 Fifth Floor Plan

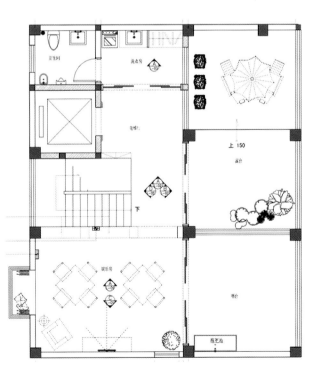

六层平面图 Sixth Floor Plan

海天别墅，岛上的隐逸假期
VILLA EMBRACED BY SKY AND SEA, LEISURELY VACATION ON THE ISLAND

- 项目名称：马尔代夫One&Only美丽岛
- 设计公司：邓尼斯顿国际建筑师与规划师事务所
- 建筑师：Jean-Michel Gathy
- Project Name: One&Only Reethi Rah
- Design Company: Denniston International Architects and Planners
- Architect: Jean-Michel Gathy

幽静、华丽、广阔，正是马尔代夫One&Only Reethi Rah度假村的最佳写照。One&Only Reethi Rah坐落于马尔代夫最大的度假村岛Reethi Rah上，是独一无二的全别墅式度假胜地。在当地的迪维希语中，Reethi Rah意为"美丽的岛屿"。度假村由印度洋上130幢最华丽、最开阔的别墅组成，从带有宽敞露台及直通海洋的小径的华丽海滩别墅，到设有大型日光浴平台、海上吊床及充满休闲气氛的尊贵水上别墅，一一演绎了终极的奢华体验。海滩别墅坐落于占地440 000 m²的热带园林内，其中多幢更设有私家泳池。

宾客的隐私是度假村考虑的重点。度假村内的海滩别墅皆相距约20 m之远，水上别墅则以每四间为一组，并通过各自的小径与岛湖相连，使度假村更添私密性。

由拱形竹板构成的高耸透气的天花令别墅内洋溢着高贵的时尚氛围。室内大量采用天然物料，如椰壳、海草、丝绸、藤枝、柚木、红木、水磨石等，与屋内的简洁线条相得益彰。

度假村内有54幢海滩别墅，占地达135 m²，全部坐落于独立的滩头之上，使这里壮阔的海洋景色一览无遗。大型的露台拥有风景如画的户外休憩区，可供宾客闲坐或用饭，并配备室外淋浴间，让宾客俨然置身于世外桃源之中。而其中27幢泳池海滩别墅均设有户外私人泳池。此外，度假村内30幢水上别墅设有分层式露台，并附设大型网状海上吊床，让宾客尽情享受微风轻抚下的日光浴或观赏宁静的闪烁星空。

2幢尊贵水上别墅及5幢尊贵海滩别墅则为宾客提供更大面积的住宿空间。水上别墅坐东向西，拥有迷人的马尔代夫日落美景。每间别墅内均附设可俯瞰海景的大型私人泳池、阔大的半开敞式平台、日光浴平台以及设于露台边缘的网状海上吊床，室内更设有独立餐厅。而尊贵海滩别墅坐拥自家海滩地段、独立泳池、遮阳日光床及独立餐厅，堪称豪华度假村的终极演绎。

来自马来西亚邓尼斯顿国际建筑师与规划师事务所的得奖建筑师Jean-Michel Gathy表示："我希望房间设计简洁，在简单长方形的布局中渗出丝丝震撼感觉，并且充满一种奢侈得近乎夸张的豪华空间感。我们希望别墅富都市气息之余，亦不失小岛特色。"他指出，透过运用简洁鲜艳的布艺装潢，再混合本地亚洲木材，例如柚木和椰子等，都有助于达到理想中的效果。

度假村内除了设有三间气派非凡的高级餐厅、尊贵的用餐别墅及一个独特的大厨花园外，还设有寿司吧、香槟吧、果汁吧，以及一个专业的咖啡吧。

Reethi 餐厅是One&Only位于马尔代夫Reethi Rah的全新豪华度假村One&Only Reethi Rah的主题餐厅，内部宽敞优雅，装潢美轮美奂，华美绚丽的色彩，加上四周不停变幻的海天景致，让宾客仿佛置身于梦幻剧场之中。

穿过巨型浮雕大门进入餐厅的中央区，深红的色调及云石布置显得格外夺目。而餐厅里的每一处景观设计都尽显非凡气派：擎天的漆花巨柱支撑着高耸的天花板，地面两侧各摆设两张十六座的餐桌，均是由一段5 m长的柚木雕凿而成。宾客亦可以在傍水而建的大型花岗石平台上进餐，眼前即是汪洋大海。而第三个座位区与酒窖相对，小巧别致，并设有手工雕刻的柱子、水景平台以及闪烁的意大利蓝色玻璃瓷砖。

这里是一个宁静美丽的天堂，让来宾在自然与质朴的优雅中回到心灵的原乡。

The One&Only Reethi Rah Resort in Maldives remarkably features seclusion, magnificence and vastness. Located on the largest resort island of Reethi Rah in Maldives, the One&Only Reethi Rah is a unique resort of full-villa style. Reethi Rah means "beautiful island" in the local language of Dhivehi. It consists of 130 villas, which are the most magnificent and vast in Indian Ocean, ranging from the gorgeous villas on beach with its spacious balconies and paths to the sea, to the luxurious and casual ones above water with the large sunbath platforms and hammocks, which offers an ultimately luxurious experience. The entire villa community is situated in a tropical garden, which covers an area of 440,000 m^2, and many of them have private swimming pools.

The resort has laid emphasis on the privacy of guests. The distance between each villa on the beach is about 20 m, while the villas above water are divided into groups of four, each of which is connected to the lagoon with its paths, which adds a sense of privacy to the resort.

Composed of the vaulted bamboo plates, the towering and well-ventilated ceiling creates an elegant and fashionable atmosphere in the villa. With a lot of natural material, such as coir, sea grass, silk, rattan, teak, mahogany and terrazzo, the design of its interior matches well with the concise lines in it.

There are 54 beach villas in the resort, which are 135 m^2 in area and located on separate beaches, and command spectacular and panoramic views of the ocean. The large open terraces in them have the scenic exterior resting area for relaxation and dinner, accompanied by the outdoor shower enclosures, which allows guests to feel as if they are in a paradise, while another 27 villas are equipped with the outdoor private swimming pools. Besides, 30 villas above water in the resort are built with the layered balconies and the large mesh hammocks, which makes it possible for guests to enjoy the sunbath in breeze or the tranquil and starry sky.

Two luxurious villas above water and five beach villas provide more living space for guests. The former face west, so that the stunning and exclusive sunset in Maldives is available. Outside each villa there is a large private swimming pool that overlooks the sea, a vast semi-open platform, a platform for sunbath, a net hammock set on the edge of the balcony above the sea, and separate dining rooms inside. Located on their own beaches, the five beach villas exists as an ultimately luxurious resort with private swimming pools, sun-shaded beds and separate dining rooms.

As the award winner from the Denniston International Architects and Planners in Malaysia, Jean-Michel Gathy said: "I want to make it impressive and create a sense of special extravagance and luxury in a simple rectangular layout, while maintaining the simplicity. Villas should present urban fashion while keeping the characteristics of the island." He pointed out that bright and fabric décor, as well as the Asian local timber, such as teak and coconut could help to reach an ideal effect.

Besides the three majestic and elegant restaurants, a noble dining villa, and a unique chef garden, there are sushi bars, champagne bars, juice bars, and a special bar for coffee.

Reethi Restaurant is a theme dining space. Spacious and elegant inside, it is beautifully decorated with a large range of brilliant colors. The shifting views of the sea and sky make guests feel as if they are in a fantastic theater.

The centre area of the restaurant behind the huge door with sculpted reliefs is in a particularly striking crimson hue with the marble furnishings. The towering ceiling is supported by the soaring painted pillars with the floral patterns. Two tables with sixteen seats for each are placed on the sides, each is made of a piece of five-meter-long teak. Guests are allowed to have dinner on the large granite platforms next to the sea and enjoy the view of the ocean. The third sitting area is exquisite and situated opposite the cellar, with the hand-made pillars, a platform for waterscape and the shiny blue Italian glass tiles.

It is indeed a beautiful and tranquil heaven for guests to return to their spiritual homeland of natural and simple elegance.

热带花园，坐享印度洋海景
ENJOYING VIEWS OF THE INDIAN OCEAN IN A TROPICAL GARDEN

项目名称：巴厘岛勒吉安酒店
Project Name: The Legian Bali

勒吉安酒店是独家经营的全套房酒店，掩映在热带花园的景观之中，尽享印度洋无与伦比的壮阔海景。宁静的海岸在其一侧相伴，这座豪华府苑带给人们一个难得的机会，让人们在这个轻松的环境里体验到巴厘岛的神奇之处。

酒店中花园的景观与清幽的荷池融为一体，美轮美奂。它拥有67间精致的套房，风格各异，令人眼花缭乱，带给客人们顶级的度假乐趣，其间有宽敞的卧室、浴室、独立的起居生活区和私人阳台。

35间单卧室豪华套房，彰显豪华舒适。宽敞的外部空间，极宜享受日光浴。大尺寸的起居空间、设施齐全的餐饮空间，提升着空间的娱乐品位。

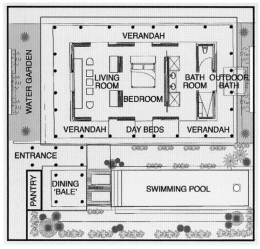

平面布置图 Layout Plan

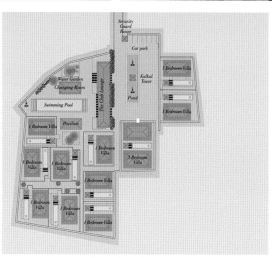

场地平面图 Site Plan

The Legian is an exclusive all-suite hotel set in landscaped tropical gardens, offering unrivalled views across the Indian Ocean. Flanked by a quiet beach, this deluxe property offers guests a unique chance to experience the magical island of Bali in a relaxed setting.

Beautifully integrated within landscaped gardens and lotus ponds, the Legian boasts 67 suites in an ample variety of exquisite accommodation styles, offering guests the finest pleasures of resort living with spacious bedrooms, bathrooms, separate living areas and private balconies.

The appeal of the 35 One-bedroom Deluxe Suites is their apparent likeness to cozy luxurious apartments. These suites boast large exteriors perfectly for sunbathing and sizeable living spaces with well-appointed dining areas for stylish en-suite entertaining.

凭栏处，自然风景如画
COMMANDING A FINE VIEW OF LANDSCAPE

◆ 项目名称：金刚山安那堤大酒店
◆ Project Name: The Ananti

安那堤大酒店位于金刚山旅游区，在金刚山脚下，四周是原始森林、起伏的群山、温泉和壮观的瀑布。

壮丽的半岛景色，如画的金刚山风景，这里是韩国最美的景区之一，可以俯瞰浩瀚的东海。

所有的客房里均可饱览高尔夫球场和山势的全景。每间客房都有宽敞的浴室，内有步入式温泉浴缸和独立淋浴区，所有客房和套房内都摆设着定制家具。

低层客房附设有平台，高层的则设有阳台。套房的休息室开放宽敞，内设舒适宜人的用餐空间和阳台。

Surrounded by pristine forests, rolling hills, hot springs and dramatic waterfalls, the Ananti is located in the Mount Kumgang Tourist district and situated at the foot of Kumgang Mountain.

Boasting stunning views of the peninsula and surrounded by the picturesque mountain scenery of Kumgangsan, this is one of the most beautiful areas in Korea, overlooking the East Sea.

All rooms boast a panoramic view of the golf course and the mountainous terrain. Spacious bathroom in each room has a walk-in hot spring bathtub and a separate shower area. All rooms and suites are furnished with custom-made furniture.

There are terraces in lower level rooms and balconies in upper level ones. The lounges in the suites are large and open with comfortable dining spaces and balconies.

本土风情，融入时尚设计
THE LOCAL ELEMENTS INTEGRATED INTO THE FASHIONABLE DESIGN

- 项目名称：巴厘岛 W 度假村
- 设计公司：AB Concept
- 面积：31 000 m²
- Project Name: W Retreat & Spa Bali
- Design Company: AB Concept
- Area: 31,000 m²

AB Concept 是一家立足于香港的顶尖室内设计公司。位于巴厘岛的时尚度假式酒店 W 度假村是其又一力作，势将引领度假酒店设计的最新潮流。为配合酒店一贯大胆创新的路线，AB Concept 在巴厘岛的传统设计中融入了崭新概念，制造惊喜。设计师将鲜明的色彩、独特的建筑轮廓和传统的设计相结合，营造出一个高品质、高格调的时尚空间。

W 度假村坐落于巴厘岛著名的 Seminyak 海滩，是岛上顶级餐厅、艺术廊及精品店云集的潮流胜地。酒店楼高四层，占地 31 000 m²，设有 232

间客房，以及79间拥有1至3间客房的独立别墅，尽显豪华气派，也令整个地区倍添色彩。

AB Concept 创办人及总监伍仲匡表示："W 度假村项目冲破了传统酒店设计的局限，融合了巴厘岛的本地元素，缔造出巴厘岛独一无二的酒店形象。"

酒店的室内设计也由 AB Concept 一手包办，焦点是令人惊艳的酒店大堂和开放式 W 休息室的设计。设计师利用巨型沙发床和飘逸的窗帘营造出雅致写意的氛围，而夸张的轮廓加上夺目的粉红色和紫色的灯光，则尽显其大胆创意。设计师采用巴厘岛常见的柚木和水磨石作为建筑材料，在设计上体现了对当地传统文化的崇敬之意。

开放式的大堂一年中大部分时间都有自然风流通，同时模糊了室内和室外的界限。人造藤条织成的灯罩仿佛是宝塔的顶端，水磨砖切割的纹理则像叶脉一样，配合镶嵌于内的贝壳，闪闪发亮。酒店的规划从巴厘岛上的日落和海景中汲取灵感，充分利用海岸景观。从大堂紫罗兰色调的灯光，到客房的水绿色，AB Concept 在设计中天衣无缝地融入了自然的元素。

AB Concept, a leading Hong Kong-based design and architectural studio, has completed its newest project, a stylish resort hotel W Retreat & Spa Bali – Seminyak, which is bound to lead the newest trend of resort hotel design. AB Concept offers a surprise to people with new concepts integrated into the traditional design of Bali to be consistent with the continuous innovation of the hotel. With brilliant colors, distinctive contour of the hotel and the traditional style, the designers create a high-quality and elegant fashionable space.

Located along Bali's renowned Seminyak beach, the hotel is a home for the island's best restaurants, galleries and designer boutiques. Comprising 232 guest rooms and 79 one, two or three-bedroom private villas, W Retreat & Spa Bali covers an area of 31,000 m^2 over four floors, which highlights its splendor and complements the whole area.

"With the W Retreat & Spa Bali we are pushing boundaries for Resort hotel design, creating a distinctly unique look that still incorporates Balinese elements." explained AB Concept cofounder and Director Ed Ng.

AB Concept designed the interiors of this beachfront resort, including the stunning terraced lobby and the open W Lounge. Oversized day beds and flowing curtains bring out an elegant and relaxing ambience, while the quirky contours and a palette of hot pinks and purples hint at the designers' adventurous spirit. While using the teak-and-terrazzo formula so popular in many of Bali's resorts, AB Concept remains respectful of the island's rich cultural heritage.

The open lobby provides natural ventilation for much of the year and blurs the boundaries between indoors and outdoors. Lampshades woven from synthetic rattan resemble pagoda roofs while the texture of terazzo tiling which has been cut resembles leaf veins and shimmers with inlaid seashells. Inspired by the colors of Bali's sunsets and seascapes, the scheme pulls in the coastal landscape at every opportunity. From the violet-hued luster of the lobby to the aquatic greens of the guest retreats, AB Concept has seamlessly integrated natural elements into the design.

传统与现代的无缝对接
THE SEAMLESS CONNECTION BETWEEN TRADITION AND MODERNITY

◆ 项目名称：印度马拉巴大酒店
◆ Project Name: The Malabar Hotel

科钦堡历史悠久，名扬天下。科钦堡的核心地理位置能为本案拥有，不得不说是别具殊荣。本案正对圣弗朗西斯教堂，观光、购物、港口游轮旅游极为便利。3 000 m之外，犹太人独家经营的古董、古玩市镇令人大开眼界。距闻名遐迩的文伯纳德湖，也仅仅一个小时的车程。

与其说本案是科钦的一个酒店，倒不如说是科钦堡和喀拉拉邦记载面料纺织的历史博物馆。其历史源远流长，可以追溯到1755年。当年荷兰人赫尔曼从马修·亨里奇手中购得此处房产。随着历史的沉淀，几经香料商人、茶叶商和银行家转手，本案在1996年成为科钦堡的第一家精品文物酒店。

酒店装饰考究，精心挑选的古董艺术品彰显细部精华。传统与现代实现完美无缝对接。具有深厚历史文化底蕴的港口城市生机无限。一个空间有限的人间天堂，竟然不可思议地在这里呈现。空间内，文物承担起传承文化的重担。艺术在这里起到更好的作用，衬托着空间气韵的典雅与高贵。喀拉拉邦是连接东西方之间的通道，它的精神在本案酒店空间里得到了更好的体现。

天堂一般的酒店共有17间标准客房、套房：一楼1套复式套房、5套屋顶花园套房，二楼11间豪华客房。

It is a privilege for the Malabar Hotel to be located in the heart of the historical and world-famous city of Fort Cochin. Opposite the St. Francis Church, it is ideally located for sightseeing, shopping and cruise travels. 3,000 m away, the town would feast your eyes with its curios and antiques exclusively managed by the Jews. The well-known Vembanad Lake is just an hour's drive away.

The Malabar Hotel is more of a historical museum that records the fabric history of Fort Cochin and Kerala than a hotel in Cochin. It dates back to 1775 when Jan Herman Clausing, a Dutch, bought the property from Mathew Henrich Beyls. As time went by, it was subsequently owned by spice traders, tea traders and bankers. In 1996, it became Fort Cochin's first boutique heritage hotel.

Decorated with the strictly-selected artistic antiques designed to the last detail, the Malabar Hotel seamlessly combines tradition and modernity. With a deep historical and cultural background, the harbour city is animate and energetic. It appears here as an amazing heaven with limited space. In the Malabar Hotel, the cultural relics shoulder the responsibility of inheriting traditional culture. The artistic designs function as the background of an elegant atmosphere in the hotel. The Malabar Hotel is a passage between the East and the West, which is best represented by this hotel.

The Malabar Hotel has 17 standard rooms and suites in total: a duplex Malabar suite and 5 roof garden suites on the first floor, and 11 deluxe rooms on the second floor.

240

MYSTERIOUS ARAB STYLE

神秘阿拉伯

丝路繁锦，古都神韵
THE SILK ROAD, THE ANCIENT CHARM

项目名称：西安金地湖城大境
设计公司：广州市韦格斯杨设计有限公司
设计师：区伟勤
面积：390 m²

Project Name: Golden Land & Lake City
Design Company: GrandGhostCanyon Designers Associates Ltd.
Designer: Ou Weiqin
Area: 390 m²

本案共四房两厅四卫，分地上和地下层，地上约250 m²，地下约140 m²。整体包括三重庭院，其总面积约200 m²。另有双层挑高客厅、大面积转角露台以及半地下室情境私享处。

设计师利用新中东风格的设计泛起奢华浪花，华贵与惊艳并融。可以想象中东异域那引人入胜的万种风情，沿着时光的隧道，带你进入一个如梦似幻的丝绸世界，使本案凸显亚洲东方文化神韵。

首层公共领域置以开放式的大尺度空间设计理念，波斯蓝、月光白的主色调贯穿其中，古朴而沉实的家具，呼应"平滑空间"的概念。没有封闭的空间里可以看见边界，空间语言的创新伴随着身体漫游而进行。庭院、阳台以或动或静的不同风格区分，室内以双层挑高等元素统筹，让内部空间流淌着自由、从容的气氛，与中东传奇的壮丽变幻不谋而合。

起居室是5.7 m开间，外连南北广阔庭院，内为双层挑高客厅，气派轩昂。你可以感受到远古东方的瑰丽风格荡漾于宏伟的景观内。从伊朗购得的波斯宫廷软塌沙发位于客厅中间，墙上挂有关于迪拜骆驼、沙漠晨曦、希腊阿帕农神殿、阿拉伯的穹顶等多个主题的旅行摄影作品，新贵的奢华浓厚的异国情怀扑面而来，空间内弥漫着浪漫的人文气息，返璞归真的田园风情独具亲和力，是品质生活的意志表达。

"葡萄美酒夜光杯"，将岁月的爱酿成醇酒。视线触及之处，酒吧台惹人酒意。巨幅羊皮卷地图撩起人们畅游异域的雄心，恍如隔世，而在甘醇的琼浆里游走又是另一番滋味。

餐厅里展示着好客主人的收藏。餐椅为新古董风格装饰艺术，更添韵味。餐桌有简洁的桌面和独具韵味的桌脚，餐桌上摆放的餐具，其精致的纹路、华丽的色彩令人赏心悦目。

拐个弯，走进二层私人空间，这里别有一番风趣。主卧豪华套房以独立卫浴及步入式衣帽间为起点，开阔的空中花园观景以开放式思维设计，视线可以从一个点跳到另一个点，不会错过春天的花开。主卧以中东宫廷风格神韵打造，奢华和浪漫相融。

接下来是书房。主人酷爱阅读和旅行摄影，书房里收藏了从中东搜罗的工艺品，极富造型感，精致夺目。和书生气息不同的是浪漫情怀，且看隔壁的冲晒室，这里有主人收藏的各式型号的相机。冲晒室不但显示了主人的特殊爱好，其照片亦记录了旅途的无限风光和迷人故事。

或悲或喜，独处可以尽情欢饮。这里是地下室情境私享处。它的独特之处在于以四韵为主题的雅士生活：花韵、茶韵、音韵、水韵。

花韵：奢华的下沉式庭院，繁花处处。柔软的阳光恣意倾泻于金漆台柜、镀金器皿等伊斯兰家具上，高贵传统一目了然，化繁为简的现代笔触让人心神荡漾和迷醉。

茶韵：茶香人醒，饮茶经由丝绸之路传到西域各国，中东品茶的风味以浓郁迷香见称。从西域搜罗的各色茶具巧置茶室，得来不易，异域香

茗飘香四溢。

音韵：屏风和民族乐器相邀以助雅兴。中东民族乐器音色婉转悠扬，细心聆听又有瑰丽传奇。

水韵：何以俘获女人心？且看私人水疗室。香薰带着阿拉伯劳伦斯的浪漫故事气息，音乐节奏舒缓撩人。整个地下室私享处凸现出亚洲东方文化神韵，使得瑰丽的中东风格一览无遗。

The two-storey villa comprises four bedrooms, one living room, one dining room and four bathrooms. The overground floors have an area of 250 m^2 while the underground floor about 140 m^2. The villa has three courtyards, which cover an area of 200 m^2 in total. In addition, there is a double-height living room, a large terrace with a turning corner and a semi-underground private space.

The neo-Mideastern style in the villa is luxurious and stunning. You can imagine that the fascinating, tempting and exotic Mideastern flavour there leads you to back through the time tunnel to a fantastic world of silk, which highlights the cultural charm of East Asia.

The design concept of a large open space is applied to the public area of the first floor, which is mainly painted in Persian blue and moonlight white. The furnishings are primitive, simple, heavy and solid to meet the concept of the "smooth space". This open space has clear boundaries. As you tour around the villa, you could feel the innovation of the design. The garden and the balcony are either in a static or in a dynamic style, creating a free, relaxing and leisurely atmosphere along with the element of the double-height stories, which is consistent with the changeable magnificence of the legendary Mideast.

The 5.7-meter-high living room is connected with the wide courtyard outside, and has a majestic double-height space inside. You can feel the ancient oriental magnificence in this spectacular architecture. Besides, with the soft Persian palace couch bought from Iran and the photographs taken in the trips on the wall about Dubai Camel, Desert Dawn, Greek Apanon Temple and Arab Dome, the living room presents a deep exotic and luxurious feeling of the emerging high social class. The space is filled with a romantic and cultural atmosphere. The original pastoral style with affinity expresses the yearn for a high-quality life.

As an ancient poem suggests, it is the good wine made from the everlasting love that is enjoyed by people in the moonlight. As far as your eyes could reach, the bar counter would arouse your desire for wine. People will be motivated to explore the exotic lands by the large scroll map here, as if being in another world, while they are allowed to feel a new flavor when enjoying the good wine.

The dining room shows the owner's collection. The chairs are of neo-classical style of Art Deco. The table has a simple desktop, unique desk legs and tableware with exquisite textures.

Around the corner is the private space of another tone on the second floor. The deluxe master suite is equipped with a separate bathroom, a walk-in closet, and an open hanging garden. The space is inspired by divergent thinking, which allows your sight to skip from one place to another so that you won't miss anything. The master suite is designed in the luxurious and romantic style of the Mideastern palace.

In the study, the owner, who is passionate for reading, traveling and photographing, displays a collection of unique and exquisite artifacts from his travels to the Mideast. Except for the cultural atmosphere in the study, a romantic tone can be felt from the film studio next door. There is also a collection of cameras of different types, which shows not only the owner's special hobbies, but also the beautiful scenery and fascinating stories in the journey recorded in the photos there.

You can drink as much as you can, indulging yourself in various moods in the private space in the basement, which features four themes of gentleman's life: flowers, tea, music and water.

Flowers: a generous sunken garden is planted with flowers everywhere. Sunlight spills on the Islamic furniture like gold lacquer cabinet and the gilded utensils, which are in the style of noble tradition and the fascinating modernity that reduces the original complex structure to the simple layout.

Tea: the fragrance of tea is stimulating. Tea culture was spread to the Mid-Asian countries through the Silk Road. In the Mideastern countries, tea characterizes its intoxicating aroma. The tea room is placed with the precious tea sets from those countries, releasing exotic tea fragrance.

Music: the screen and ethnic musical instruments help to cultivate a noble and refined sense. The Mideastern ethnic musical instruments sound melodious and legendary.

Water: how to win a woman's heart? Here comes the exclusive private spa room, which is filled with a romantic atmosphere from the story of Lawrence of Arabia in a slow and seductive musical rhythm. Meanwhile, the private space in the basement highlights the essence of East Asian culture and displays the startling Mideastern style.

豪华之巅，尊贵无限
THE UTMOST LUXURY, THE EXTREME NOBLENESS

◆ 项目名称：阿布扎比"豪华之巅"别墅
◆ Project Name: Nurai Resort Island

小岛Nurai名字源于阿拉伯语"nour"，意喻"光"，它是阿布扎比最新开发项目之一，营造超级富豪的梦幻天堂。该岛可谓Zaya的开山之作。Zaya总部位于阿联酋，以精品豪华房产开发为己任。这座私人岛屿开工于2008年5月，竣工日期定于2011年12月，距阿布扎比陆地咫尺之遥。这里充满度假风情，乃绝世之作，距离城市枢纽如萨迪亚特和亚斯岛商业区只有几分钟路程，生活便利，尽享世界顶级娱乐、文化、零售、餐饮、教育和医疗服务设施。

《新闻周刊》独具慧眼，特授其"世界上最豪华项目"之荣誉。Nurai独家推出私人豪华岛上度假区，包括限量版的28幢海滨别墅，12幢水上别墅和一处精品豪华静居处。世界上最知名的建筑师、设计师和设计品牌公司倾情合作，参与到Nurai的发展中来，其中包括闻名遐迩的建筑和设计公司AW2建筑事务所。

Nurai，一个终极的天堂，这里环绕着水晶般清澈的海水和天然纯净的美丽沙滩，如一串项链，熠熠发光。水是Nurai的主题，漫步其间，你可以感受到它的无限魅力。水上别墅延伸出海，如临波踏步；海滨别墅，临海岸峭壁而立。树木茂密，风拂屋顶，万顷碧涛，如天然华盖，掩映着别墅。原生态的海天美景顷刻间与滨水美学合二为一。从别墅处俯瞰，是碧波、沙滩、绿树和灌木，栋栋建筑如同遁形。身居内里空间，放眼望去，窗外风景一览无遗，动人心魄。大部分房间有挑高的3 m天花板，中庭更是高达7 m。壮观的空间两侧由镶板簇拥，锦上添花，在现代设计的基础上，融入阿拉伯的风情，给人深刻的印象。整体空间采光自然，通风顺畅。静卧于主卧之中，可俯瞰壮阔的中庭风景。

绿毯一般的建筑表皮覆盖着别墅整体结构，这里有居住者私人独享的奢华世界，以及舒适宜人的海滨别墅空间，从1 645.03 m^2至1 740.07 m^2不等，室外地基从1 950.96 m^2至11 055.46 m^2各异，还有大面积的室内外空间，包括无边际泳池、户外烧烤区、私人海滩、花园、娱乐庭院、厨房及隐蔽服务区域。

度假区的水上别墅提供无与伦比的水上生活体验，室内面积930.05 m^2，室外从1 950.96 m^2到4 738.05 m^2大小不等。这些别墅也从豪华静居处所提供的设施和服务中得到便利，别具特色，拥有360°大景观阳台、落地玻璃窗、屋顶休息室和花园、无边际泳池、复合平台和烧烤区，适合于全户外用餐或一边晒太阳，一边饱览阿布扎比海洋和天际线的全景。服务区域隐蔽式设施及最新技术使生活变得如此简单。

除每栋别墅内部各有特点外,居住者们还可以享受到小岛豪华精品静居处所提供的便利设施和服务,静居处为居住者带来全方位的隐私,有浮动的小艇停泊区,内有为居住者和客人接风洗尘的休息室、酒店服务、私人直升机停机坪、远近驰名的豪华水疗中心、世界一流的健身中心、众多世界级的餐馆和休息室,包括该地区首家Cafe Del Mar咖啡馆和Sir Alan Yau(米其林星级厨师)餐馆,供岛上居住者和客人尊享。

地处阿布扎比东北,这座美丽的岛屿远离城市的喧嚣,离陆地只有几分钟路程,对居住在其中的人们来说如同通往阿布扎比的世外桃源,它是一个有影响力的、不断增长的商业枢纽,也是一个新兴的文化和娱乐热点。

Derived from the Arabic word "nour" meaning "light", the island of Nurai is one of the latest of Abu Dhabi's developments to catch the fancy of the ultra wealthy. Nurai is the inaugural project from Zaya, a UAE-based boutique developer specializing in exclusive luxury real estate. The private island, which was launched in May 2008, to astounding success, is due for completion in December 2011 and is located just minutes away from mainland Abu Dhabi. Nowhere in the region do residents enjoy permanent resort living while being minutes away from urban hubs such as Saadiyat and Yas Islands, which house the world's best entertainment, cultural, retail, dining, education and medical facilities.

Awarded the accolade of the "Most Luxurious Project in the World" by *Newsweek* Magazine, Nurai offers an exclusive private luxury island resort community of 28 beachfront estates of limited edition, 12 water villas and one boutique luxury retreat and is being developed in partnership with some of the world's most renowned architects, designers and brands, including the renowned architecture and design firm AW2.

The ultimate heaven, Nurai, is surrounded by crystal clear water and ringed with pristinely beautiful sandy beaches. Water is Nurai's motif and wherever you are you feel its pull. Water villas extend into the ocean while the island's beachfront estates nestle on the ocean's edge with a canopy of undulating greenery unfolding across the roof of the estates to give the illusion of an untouched virgin island landscape coupled with modern beachside aesthetics. Viewed from above the estates practically disappear from sight while the sea, the beach, the trees and shrubs can be seen. Every window of every room frames a new stirring panorama. Most rooms have generous three-meter-high ceilings, and the central atrium is 7 m in height. Musharabiya panels flank this spectacular void, adding to its presence and introducing a dramatic Arabesque touch to what is predominantly a contemporary design. The overall effect is light and airy, with the master bedroom overlooking this grand central space.

The sweeping structures underneath the "green carpet" envelop the residents in their own world of private luxury, with livable space in the beachfront estates ranging from 1,645.03 m^2 to 1,740.07 m^2, outdoor plots ranging from 1,950.96 m^2 to 11,055.46 m^2, and expansive outdoor and indoor spaces including infinite pools, outdoor barbeque areas, private beach and gardens, entertainment patios, chef and show kitchens and concealed service quarters.

The resort's water villas offer an unparalleled over-water living experience, covering an area of 930.05 m^2 inside, and outdoor plot of 1,950.96 m^2 to 4,738.05 m^2. The villas, which will also benefit from the amenities and services provided by the luxury retreat, are truly unique with their 360° terraces, floor to ceiling windows and roof top lounge/garden, infinite pool, multiple decks and barbeque area suitable for "al fresco" dining or simply lazing in the sun whilst enjoying the panoramic views of the Ocean and Abu Dhabi sky line. Concealed service quarters and the latest technology allow for an easy life.

In addition to the in-house features of each of its villas, residents will have access to the services and amenities provided by the island's luxury boutique retreat (strategically positioned at the island's tip to afford residents complete privacy), which include a floating marina with arrivals lounge for residents and guests, service access from the hotel, a private helipad, a celebrated luxury spa, a world class fitness center, and a multitude of world-class restaurants and lounges including the region's first ever Cafe Del Mar and a Sir Alan Yau (Michelin Star Chef) inspired restaurant exclusively for residents of the island and guests.

Lying immediately northeast of Abu Dhabi, the beautiful island is set in isolation away from the hustle and bustle of city life yet within minutes from the mainland, affording its residents secluded accessibility to Abu Dhabi, an influential and growing business hub and emerging cultural and entertainment hotspot.

糅合奇幻梦想与不朽传统
A MIX OF FANTASY AND MONUMENTAL TRADITION

◆ 项目名称：迪拜One&Only Royal Mirage度假村
◆ Project Name: One&Only Royal Mirage

迪拜One&Only Royal Mirage度假村的Arabian Court 乃潜藏于隐蔽绿洲里的古老大宅，其独有的内在美态深藏不露，乍看起来，既神秘又迷幻。Arabian Court位处优雅喷泉、蜿蜒绕道及茂密园林之间，其瑰丽建筑均为雕栏玉砌，展现对称美学，同时反映东方工艺精髓。当宾客亲临其中，走访多个怡人庭园之际，将可从流水到奇石，从奇石到巧木，从棕榈树到碧海，逐一细赏Arabian Court的真貌。

宾客在抵达Arabian Court时，迎面即见波平如镜的水池。水池泛起闪闪波光，宾客放眼远眺水平线，即可瞥见庭园背靠的一片碧海。入口设计精巧，棱角分明，宾客通道是一条气派非凡的画廊。画廊是Arabian Court的重要枢纽，分别连接两翼大楼，引领宾客前往162间温馨精致的客房，以及10间风格独特的套房。所有房间均设有私人露台，让宾客可以饱览醉人海景。

Arabian Court俯瞰长达1 000 m的私人海滩，并提供各种各样的休闲活动，包括扬帆出海、激流独木舟、风帆、滑水及钓鱼等，一应俱全。宾客亦可闲游于优美宽敞的园林中与池畔的甲板上，沉醉于和谐宁静的氛围里，享受自在闲暇。

Arabian Court亦设宏伟的阿拉伯城墙，以及能款待多达300位宾客的庄严富丽的"圆形露天剧场"，可用于举行各式特别盛事、节庆晚宴、表演及产品发布会；更配备完善的宴会及会议设施，包括Peregrine 宴会厅，能从容款待200位宾客，以及两间会议室和传统"Majlis"风格的会议酒吧。

　　Arabian Court设有三间格调迥异的时尚食府,分别是The Rotisserie、Nina及Eauzone。The Rotisserie瑰丽华美,从温馨雅致的庭院平台及开阔露台往远处看,度假村的芳香园林及优雅喷泉一览无遗;其别出心裁的开放式厨房配备高耸圆拱天花板、层层回转的铜制烟囱、搪瓷铁制烤炉及宽阔木台,令人回忆起昔日的烹厨年代。厨师会按照欧洲及阿拉伯传统烹厨方法,糅合当代技巧,精心炮制多款佳肴。期间,宾客亦可与厨师交流心得。

　　印欧食府Nina,装潢鲜艳斑斓。当宾客从楼梯拾级而下,踏进跳脱活泼又五光十色的餐厅之际,即可感觉到室内洋溢着历险探索的刺激气氛。闪烁的烛光、流丽的垂珠,映衬着强劲的音乐,合力打造奇幻醉人的氛围,让宾客于"香料佳人"Nina内度过难忘的夜晚。Nina提供极品美膳,以诱人香气触动宾客味蕾,引领他们展开探索味道之旅,发掘其中的瑰宝。

　　被蓝天碧海所环抱之Eauzone,是细观潮汐与余晖的理想地点。餐厅位处静谧的棕榈树林之间,掩映着木制甲板及漂浮的"Majlis"用膳厅,宾客在此可饱览棕榈岛海湾及清澈泳池的景致。夜幕低垂,精致流丽的Eauzone散发动人魅力,呈献糅合亚洲风味的当代特色佳肴。

　　天台酒吧及露台乃度假村最为极致的室外夜店,营造融入不同元素及阿拉伯特色的餐饮体验。当宾客步入其中时,即见光彩夺目且舒适时尚的圆形吧台,并能安坐其中,观赏电视上不同的运动节目。室内弥漫着轻松闲逸的氛围,当宾客走过楼梯时,闲晃于布满星辰的拱顶天花板之下时,亦能享受到当中的无尽诗意。

　　Samovar酒吧富丽堂皇,设有高耸的天花板,与大堂气派非凡的画室互相辉映;室内摆放着巧夺天工的手制家具,缀以金叶拱顶及奢华露台,在此宾客可全面饱览园林泳池及波斯湾的迷人风光。日间及傍晚时分,酒吧均提供多款精致糕点、高级下午茶、特色咖啡及简单小吃。

The resort Arabian Court in One&Only Royal Mirage is an ancient mansion embedded in an oasis. Its beauty is hidden inside, appearing mysterious and magical at first sight. Situated among the beautiful fountains, winding paths and thick forests, the Arabian Court reveals symmetrical aesthetics, reflecting the essence of the oriental artifact with its carved railings and jade inlays. Guests are allowed to enjoy the view of the Arabian Court from streams to stones, from stones to woods, from palm trees to the sea while paying a visit to the beautiful gardens.

When guests arrive at the Arabian Court, they'll catch sight of a placid pool. With the sunlight reflected on the surface of the pool, guests would get a glimpse of the sea at the back of the courtyard if they overlook the horizon. The entrance is exquisite and angular. The splendid gallery for guests serves as the backbone that connects two wings, and leads to 162 warm and sophisticated guest rooms and 10 unique suites, all of which are equipped with private terraces for the stunning sea views.

The Arabian Court overlooks the 1-kilometer-long private beach and offers varieties of leisure activities, such as sailing, whitewater canoeing, windsurfing, water skiing and fishing. Guests could also wander in the beautiful and spacious gardens and on the decks of pool, indulging in the harmonious and peaceful atmosphere and enjoying a moment of idleness.

In the Arabian Court there is a grand Arabian wall and a magnificent amphitheater that can accommodate 300 guests for various special events, festival parties, shows and product launches. Perfectly equipped facilities for banquets and conferences include Peregrine, a banquet hall that can entertain 200 guests, two conference rooms and one traditional conference bar of "Majlis" style.

There are three fashionable restaurants of different styles, namely the Rotisserie, Nina and Eauzone. The Rotisserie features resplendence, with its warm and delicate platform in courtyard and the open and broad terrace, both of which command a fine view of the fragrant gardens and delicate fountains. The unique open kitchen accompanied by a towering arched ceiling, the copper chimney that winds around and around in circles, the ceramic and iron oven and the wide wooden tables combine to remind us of the old days when people cook in traditional ways. Chefs are experienced in combining traditional European and Arabian cooking methods together with modern techniques to make delicacy. Meanwhile, exchanges of cooking experience can take place between guests and chefs.

Nina, an Indo-European restaurant, is decorated gorgeously. Guests could enjoy a stimulating and adventurous atmosphere when they enter the lively and animated restaurant from the stairs. With flickering candlelight, hanging beads and strong background music, the fantastic and intoxicating ambience allows guests to spend a memorable night. Nina offers attractive and fragrant gourmet, leading to the exploration of flavors and discovery of treasures.

As for Eauzone, embraced by the blue sky and green sea, it's really an ideal place to appreciate the tides and sunset. It is located in the palm trees, which offers a shelter for the wooden deck and the floating "Majlis" dining room, where guests can enjoy the view of the palm island bay and the clear swimming pool water. As night falls, the Eauzone becomes more fascinating, presenting the special cuisine blending contemporary and traditional Asian styles.

The bar and the terrace on the rooftop are the most remarkable nightclub in this resort, where the dining experience is mingled with different elements and Arab styles. When guests stepped into the club, they could catch sight of the bright, comfortable and stylish round bar, at which they can sit comfortably and watch different sports programs on TV. In such a leisurely and cozy atmosphere, guests can feel endless poetry that transcends the vaulted starry ceiling.

In the majestic Samovar Bar are the soaring ceilings echoing with the fantastic studio in the lobby. Inside the room are extremely delicate hand-made furniture, a dome inlaid with gold leaves and a luxurious terrace where guests can enjoy the panoramic view of the swimming pool and the Persian Gulf. A variety of fine pastries, afternoon teas, specialty coffees and light snacks are available in the day and in the evening.

281

283

异域风情，圣洁宁静
THE EXOTIC THEME, THE HOLY AND TRANQUIL ATMOSPHERE

项目所在地：阿联酋
Location: The United Arab Emirates

本案坐落于阿联酋的首都阿布扎比，1996年开始施工，历时10多载，耗资25亿多美元。

本案占地26 000 m^2，整个建筑群都镶嵌着全白色大理石，均是来自意大利、希腊、印度和中国的28个不同品种的大理石，因此目光所及之处无不让人眼花缭乱。纯净而剔透的白色象征着和平与友爱。

4个高耸入云的尖塔是本案的一大特征，每个高107 m，并由57个大小不一的大理石穹顶组成，这些穹顶分布在各个入口。院子由上等的印花大理石铺成，面积达17 000 m^2。本案已经成为阿布扎比的标志性建筑之一。

整栋建筑选用了来自世界各地的昂贵原材料和艺术装饰品，错综复杂的雕刻和壁画由各个国家的艺术家共同创作完成。这里铺设着世界最大的手工编织地毯，悬挂着价值800多万美元的、由德国制造的世界第一的镀金枝形吊灯。三个大圆顶下的广阔空间可容纳超过6 000人，整栋建筑则可容纳40 000多人。本案拥有多项世界之最，其中之一就是这里摆放着的世界上最大的手织波斯地毯，整块地毯面积达5 627 m^2，材料都是来自伊朗和新西兰的顶级羊绒，由1 200多名女性手工编织者、20名技术人员和30名工人历时18个月编制而成的。

另外，各种吊灯把整栋建筑点缀得熠熠生辉。最大的一个水晶吊灯直径10 m，高达15 m，重达10 t，令人震撼。

雪白的大理石圆顶和墙面在阳光下隐隐发亮，白得一尘不染。湛蓝的一池清水营造出圣洁宁静的氛围。廊柱的柱头上，缀着镀金的椰枣树叶雕饰。整栋建筑如同一件梦幻般的艺术品，在阳光下绽放出动人心魄的魅力。

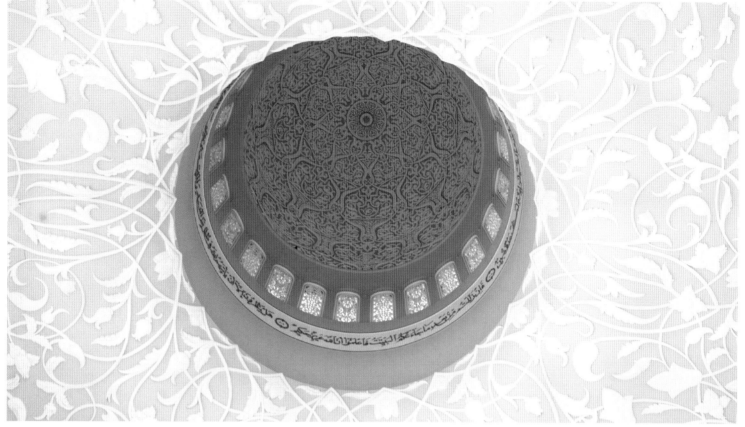

Located in Abu Dhabi, the capital of UAE, the project has been in construction for more than 10 years since 1996 and has cost over 2.5 billion dollars.

The project is 26,000 m^2 in area, and the whole complex is inlaid with white marble of 28 different kinds imported from Italy, Greece, India and China. Therefore, as far as your eyes could reach, it is dazzling and dizzying with the purity and refinement of the marble, which symbolizes peace and fraternity.

The project features its 4 towers soaring into the sky, each is 107 m, and is composed of 57 marble domes of different sizes over the entrances. The courtyard is paved with the best printing marbles, and has an area of 17,000 m^2. It has become the landmark of Abu Dhabi.

The expensive raw materials and the artistic decorations adopted by this project are selected from all over the world, and its intricate carvings and murals are created together by artists around the world. The largest handmade carpet in the world is laid here and above that the largest German gold–plated chandelier in the world is hung, which costs more than 8 million dollars. Under the three large domes, the generous space can accommodate over 6,000 people. The whole project can hold more than 40,000 people. The project has many features that are on the top of the world, and one of them is the largest hand–woven Persian carpet in the world. It covers an area of 5,627 m^2 and it is made of the top cashmere from Iran and New Zealand. It took more than 1,200 female hand–weavers, 20 technicians and 30 workers 18 months to weave the carpet.

Additionally, a variety of chandeliers decorates the interior of the project. The largest one of them is 10 m in diameter, up to 15 m in height and 10 t in weight, which is really stunning.

The white marble dome and wall of the project are shiny, bright and spotless in the sun. The clear blue water in the pond creates a holy and tranquil atmosphere. The ends of the pillars are trimmed with gold carvings of date palm trees and leaves. The whole project is like a fantastic artwork displaying its startling beauty.

瑰丽想象，沙漠绽放
THE ROSE BLOSSOMS INTO IMAGINATION IN DESERT

◆ 项目名称：多哈丽思卡尔顿酒店
◆ Project Name: Ritz Carlton Doha

阿拉伯建筑风格一向是奇想纵横，庄重而富变化，雄健而又不失雅致，中西交融，传承古今，在世界建筑之林中大放异彩。

世外桃源，清水幼沙，多哈丽思卡尔顿酒店就沉醉于远离尘嚣的碧蓝波斯湾的阳光海岛之中。其空间的气度恢弘，令人神往。无论近观还是远眺，只要你身处其中，就可悠然感受到多哈丽思卡尔顿酒店将极尽奢华的舒适典雅与阿拉伯人传统的热情好客完美地结合在一起。

多哈丽思卡尔顿酒店简直就是一座奢华大气的中东天堂，远远望去高大雄伟，气派非凡，让人不由自主地联想到埃及的金字塔。步入酒店，你可以看见各角对称的圆弧拱的入道。壮阔的大理石石柱，排列成大厅的入口。水晶吊灯的灯光流泻在石柱的背面上和空隙里，衬托着大理石地板上的石柱倒影，勾画出一幅水中画，画中景。

餐厅不仅是食客可大快朵颐的享受空间，更是一家人团聚一起交流情感的温情场所。所以一款好的餐厅设计，需要视觉上的愉悦感来营造一种温馨柔和的氛围，让身处其中的人身心得到放松。多哈丽思卡尔顿酒店共有9家不同特色的餐厅，其设计独具一格。有的餐厅全部用金色的沙雕组成，螺旋形的墙壁、镂空的拱形吊棚上晶莹剔透的水晶吊灯、角落里不经意摆放的阿拉伯小饰品，无一不让人领略到阿拉伯风格的神奇之处，富丽堂皇，巧夺天工。

374间设施齐全的客房，每一间或可独享完美海景，或可眺望高尔夫场地。客房布置主要在颜色和家具上做文章。简洁的梳妆台、原产西班牙和意大利的纯手工制作家具，简约中透露异国风情，或是湖蓝，或是深紫，甚至浅红色奢华棉麻床上用品、床帘缓缓散开，悠然垂下，曼妙的感觉引人遐想。清晨伴随着第一缕透过私人阳台的阳光，美好的一天又拉开序幕。

The Arabian architecture has always been a fantasy, solemn but changeable, vigorous yet elegant. The combination of eastern and western styles and the representation of tradition and modernity make it outstanding in the world.

It is a paradise with clean water and glittering sand. Indulging itself in the secluded Persian Gulf, the hotel on the shiny island is fascinating and magnificent. From near or far, as long as you are in the hotel, what you feel is extravagant comfort and elegance, as well as the traditional local hospitality.

The hotel appears as a luxurious and opulent Mid-eastern paradise. It's tall and grand, reminiscent of the pyramids in Egypt. With the symmetrical angles, the passage into the entrance is arched. The entrance is lined with marble pillars. The glow cast by the crystal chandelier on the back and in the intervals of the pillars, along with the shadows on the floor, makes a pictorial scene in water.

While enjoying the cuisine in the dining room, the whole family can also enjoy the happiness of family reunion. An excellent dining room requires a visual pleasure that can provide a warm and gentle atmosphere for the guests to get relaxed. The hotel has 9 dining rooms, which feature diversity and uniqueness. Some of them are wholly composed of golden sand sculptures and decorated with spiral walls. The transparent crystal chandeliers are hung from the arched hollowed-out ceilings, and the Arabian accessories are carelessly put in the corners, which reveals a wonderful, magical and magnificent Arabian style.

The 374 suites in the hotel are well-equipped, each with a perfect sea view. The golf course can also be seen from a distance. Each room features its colors and furniture, such as the simple dresser and the hand-made furniture made in Spain and Italy, which gives you an exotic feeling while maintaining simplicity. The red, blue or purple luxurious cotton bedding looks like an opening curtain, smoothly spreading out and swagging and creating a sense of fantasty. With the first ray of sunshine shining through the private balcony in the morning, another beautiful day is about to start.

图书在版编目（CIP）数据

浓浓亚洲风：东方古韵的传承与演绎 / 黄滢主编
. -- 南京：江苏科学技术出版社，2014.4
ISBN 978-7-5537-2951-0

Ⅰ. ①浓… Ⅱ. ①黄… Ⅲ. ①室内装饰设计－作品集－亚洲－现代 Ⅳ. ①TU238

中国版本图书馆CIP数据核字(2014)第046137号

浓浓亚洲风——东方古韵的传承与演绎

主　　　编	黄　滢
项 目 策 划	凤凰空间
责 任 编 辑	刘屹立
出 版 发 行	凤凰出版传媒股份有限公司 江苏科学技术出版社
出版社地址	南京市湖南路1号A楼，邮编：210009
出版社网址	http://www.pspress.cn
总 经 销	天津凤凰空间文化传媒有限公司
总经销网址	http://www.ifengspace.cn
经　　　销	全国新华书店
印　　　刷	北京建宏印刷有限公司
开　　　本	965 mm×1 270 mm　1 / 16
印　　　张	22.5
字　　　数	160 000
版　　　次	2014年4月第1版
印　　　次	2014年4月第1次印刷
标 准 书 号	ISBN 978-7-5537-2951-0
定　　　价	298.00元

图书如有印装质量问题，可随时向销售部调换（电话：022-87893668）。